Lucas D. Introna

PERSONAL COPY

New Information Technologies in Organizational Processes:

Field Studies and Theoretical Reflections on the Future of Work

IFIP - The International Federation for Information Processing

IFIP was founded in 1960 under the auspices of UNESCO, following the First World Computer Congress held in Paris the previous year. An umbrella organization for societies working in information processing, IFIP's aim is two-fold: to support information processing within its member countries and to encourage technology transfer to developing nations. As its mission statement clearly states,

IFIP's mission is to be the leading, truly international, apolitical organization which encourages and assists in the development, exploitation and application of information technology for the benefit of all people.

IFIP is a non-profitmaking organization, run almost solely by 2500 volunteers. It operates through a number of technical committees, which organize events and publications. IFIP's events range from an international congress to local seminars, but the most important are:

- The IFIP World Computer Congress, held every second year;
- open conferences;
- working conferences.

The flagship event is the IFIP World Computer Congress, at which both invited and contributed papers are presented. Contributed papers are rigorously refereed and the rejection rate is high.

As with the Congress, participation in the open conferences is open to all and papers may be invited or submitted. Again, submitted papers are stringently refereed.

The working conferences are structured differently. They are usually run by a working group and attendance is small and by invitation only. Their purpose is to create an atmosphere conducive to innovation and development. Refereeing is less rigorous and papers are subjected to extensive group discussion.

Publications arising from IFIP events vary. The papers presented at the IFIP World Computer Congress and at open conferences are published as conference proceedings, while the results of the working conferences are often published as collections of selected and edited papers.

Any national society whose primary activity is in information may apply to become a full member of IFIP, although full membership is restricted to one society per country. Full members are entitled to vote at the annual General Assembly, National societies preferring a less committed involvement may apply for associate or corresponding membership. Associate members enjoy the same benefits as full members, but without voting rights. Corresponding members are not represented in IFIP bodies. Affiliated membership is open to non-national societies, and individual and honorary membership schemes are also offered.

New Information Technologies in Organizational Processes:

Field Studies and Theoretical Reflections on the Future of Work

IFIP TC8 WG8.2 International Working Conference on New Information Technologies in Organizational Processes: Field Studies and Theoretical Reflections on the Future of Work, August 21-22, 1999, St. Louis, Missouri, USA

Edited by

Ojelanki Ngwenyama
Virginia Commonwealth University
USA

Lucas D. Introna
London School of Economics
United Kingdom

Michael D. Myers
University of Auckland
New Zealand

Janice I. DeGross
University of Minnesota
USA

KLUWER ACADEMIC PUBLISHERS
BOSTON / DORDRECHT / LONDON

Distributors for North, Central and South America:
Kluwer Academic Publishers
101 Philip Drive
Assinippi Park
Norwell, Massachusetts 02061 USA
Telephone (781) 871-6600
Fax (781) 871-6528
E-Mail <kluwer@wkap.com>

Distributors for all other countries:
Kluwer Academic Publishers Group
Distribution Centre
Post Office Box 322
3300 AH Dordrecht, THE NETHERLANDS
Telephone 31 78 6392 392
Fax 31 78 6546 474
E-Mail <services@wkap.nl>

  Electronic Services <http://www.wkap.nl>

Library of Congress Cataloging-in-Publication Data

IFIP TC8 WG8.2 International Working Conference on New Information Technologies in Organizational Processes: Field Studies and Theoretical Reflections on the Future of Work (1999 : St. Louis, Mo.)
  New information technologies in organizational processes : field studies and theoretical reflections on the future of work : IFIP TC8 WG8.2 International Working Conference on New Information Technologies in Organizational Processes: Field Studies and Theoretical Reflections on the Future of Work, August 21-22, 1999, St. Louis, Missouri, USA / edited by Ojelanki Ngwenyama ... [et al.].
      p.   cm.   — (International Federation for Information Processing ; 20)
  Includes bibliographical references.
  ISBN 0-7923-8578-0
  1. Information technology Congresses. 2. Management information systems Congresses. I. Ngwenyama, Ojelanki. II. Title. III. Series: International Federation for Information Processing (Series) ; 20.
  T58.5.I323   1999
  658.4'038—dc21   99-28492
                                                                                                         CIP

Printed on acid-free paper.

Printed in the United States of America.

Contents

Part 1: Critical Reflections

Part 2: Field Studies

Part 3: Panels

Foreword

This volume contains a collection of papers that inquire into fundamental issues of new information technologies in organizational processes. Each was reviewed by at least three referees in a double blind process. This screening resulted in a 33% acceptance rate. All the contributed papers were revised according to specific suggestions of the reviewers and program committee.

We would like to thank the General Chair, Rudy Hirschheim, the Organizing Chair, Marius Janson, and the program committee members for their hard work and dedication to making the conference a success.

Ojelanki K. Ngwenyama
Lucas D. Introna
Michael D. Myers
Janice I. DeGross

Conference Chairs

General Chair:
Rudy Hirschheim
University of Houston
U.S.A.

Co-Program Chairs:
Ojelanki Ngwenyama
Virginia Commonwealth University
U.S.A.

Lucas Introna
London School of Economics
United Kingdom

Michael D. Myers
University of Auckland
New Zealand

Organizing Chair:
Marius Janson
University of Missouri, St. Louis
U.S.A.

Program Committee

Ivan Aaen	Aalborg University, Denmark
Soon Ang	Nanyang Technological University, Singapore
Chrisanti Avgerou	London School of Economics, United Kingdom
David Avison	University of Southampton, United Kingdom
Richard Boland	Case Western Reserve University, U.S.A.
Kristin Braa	University of Oslo, Norway
Kweku-Muata Bryson	Virginia Commonwealth University, U.S.A.
Claudio Ciborra	THESEUS Institute, France,and London School of Economics, United Kingdom
Kevin Crowston	Syracuse University, U.S.A.
Liisa von Hellens	Griffith University, Australia
Matthew Jones	University of Cambridge, United Kingdom
Lynette Kvasny	Georgia State University, U.S.A.
Frank Land	London School of Economics, United Kingdom
Johnathan Liebenau	London School of Economics, United Kingdom
Kalle Lyytinen	University of Jyväskylä, Finland
Lars Mathiassen	Aalborg University, Denmark
John Mingers	University of Warwick, United Kingdom
Nathalie Mitev	London School of Economics, United Kingdom
Joe Nandhakumar	University of Southampton, United Kingdom
Steve Sawyer	Syracuse University, U.S.A.
Kylie Sayer	University of Technology, Sydney, Australia
Susan Scott	London School of Economics, United Kingdom
Graeme Shanks	Melbourne University, Australia
Julie Travis	Curtin University of Technology, Australia
Duane Truex	Georgia State University, U.S.A.
Cathy Urquhart	Sunshine Coast University, Australia
John Venable	Curtin University of Technology, Australia
Geoff Walsham	University of Cambridge, United Kingdom
Leslie Willcocks	University of Oxford, United Kingdom
Edgar Whitley	London School of Economics, United Kingdom
Chee-Sing Yap	National University of Singapore, Singapore

1 BUILDING ON A DECADE OF RESEARCH ON IT AND ORGANIZATIONS

Ojelanki K. Ngwenyama
Virginia Commonwealth University
U.S.A.

Lucas D. Introna
London School of Economics
United Kingdom

Michael D. Myers
University of Auckland
New Zealand

1. Introduction

At the close of this decade, and the "official beginning" of the new millennium, we continue to see the technologization of our social worlds, workplace, homelife and so on. In organizations, the focus of much of our research, IT, has become pervasive. The proliferation of electronic communication, collaborative, ERP, supply chain management and web technologies have all challenged our fundamental assumptions about organizations, organizing and work. Through the decade of the 1990s, IFIP W.G. 8.2 has responded to the challenges of demystifying these new information technologies and explicating their implications for everyday organizational life and activity. As a group of critically oriented scholars, we have been fortunate to engage with each other in a journey of inquiry during this last decade of the millennium that has been both exciting and rewarding. Throughout this decade, we have continued to inquire into the meaning of IT for everyday organizational life, engage in the key theoretical issues of our time and advance the state of knowledge of these new information technologies. In 1990, we began the decade of inquiry and critical debate with the Conference on Information

Systems Research: Contemporary Approaches and Emergent Traditions, Copenhagen, Denmark. In Copenhagen, we reflected upon the path that our working group (8.2) had followed, and how the traditions of research and practice that we had cultivated over the previous decade would serve us in this decade. We examined our research approaches and methods in the light of some important research questions that were emerging. In the following year, we responded to the influx of new communication and collaborative technologies into organizations, and offered a critical analysis of them in our conference on Collaborative Work, Social Communications and Information Systems, Hanasaari, Finland (1991). In Hanasaari, we turned a critical lens on the concepts of communication and collaboration underlying the technologies of the time and revealed some of their shortcomings. We had some heated debates on the organizational implications of these technologies, raised questions about their naiveté and pushed to conceptualize new forms of communication and collaboration technologies to over come their shortcomings.

By 1992, we were in the midst of the group support era, and many organizations were acquiring group technologies (or thinking about acquiring them) to support various activities from strategy planning to systems development. In 1992, we entered this discourse with our Conference on the Impact of Computer Supported Technologies in Information Systems Development, Minneapolis, Minnesota, USA. Once again, we turned our critical lens on this technology and its uses and implications for organizational activities and life. In 1993, at Noordwijkerhout, The Netherlands, we continued this discourse and examined more closely the social and organizational issues in our Conference on Information Systems Development: Human, Social and Organizational Aspects. By the middle of the decade, those almost imperceptible incremental changes in organizations were becoming more visible due to continuous waves of implementation and use of new information technologies. In 1994, at the Conference on Information Technology and New Emergent Forms of Organizations, Ann Arbor, Michigan, USA, our discourse shifted somewhat to examine more closely how organizations were changing in the context of communication and collaboration technologies. We took up some of the strands of the debate from the 1991 conference, and re-examined our understanding of how these technologies are appropriated by individuals and are implicated in the reproduction of organizational structures and forms. In 1995, we expanded the focus to work and organizing at the Conference on Information Technology and Changes in Organizational Work, Cambridge, England. Our examination of changes in work and organizing also took a turn into postmodern deconstruction of concepts of organizing, work and technology. At that juncture, we once again reopened the discourse on the meaning of technology in everyday organizational life. Then in 1997, in keeping with tradition, we once again revisited our methods of inquiry. At the Conference Information Systems and Qualitative Research, Philadelphia, Pennsylvania, USA, we critically assessed our research methods in the light of current theoretical questions and tried to hone them for future research engagements. As Klein (in these proceedings) points out, the fundamental questions of how to systematically produce knowledge in our field of study persist without definitive resolution. We, the members of IFIP W.G. 8.2, close out the decade and the millennium, once again looking ahead at emerging issues, while addressing some pressing questions of the present. At our Helsinki, Finland, conference on Information Systems: Current Issues and Future Changes (1998), we engaged in a broad and at times deeply philosophical debate about information technology and our field at the turn of millennium. In this, our final conference of the decade, we address ourselves to some pressing questions that organizations are experiencing in implementing current process-oriented information technologies such as ERP and intranets. However, we take some time to reflect on the meaning of technology in everyday life, the path that we have followed in our inquiries and approaches to generalizing the knowledge that we obtain from them.

2. Contributions

The papers in these proceedings fall into two broad categories: critical reflections and field studies of IT implementations. The critical reflections focus on theory and method in IS research. The field studies of IT implementations vary from ERP systems, through intranets and collaboration technologies to electronic commerce.

2.1 Critical Reflections

Mark Poster's keynote talk promises an interesting panoramic analysis on the impact of IT on life globally. Mark will discuss how IT is facilitating a fundamental transformation in human interaction without regard to time, geography or national boundaries. He will outline some of the political implications of global communication systems such as the Internet.

   In his insightful piece, Klein reflects on the development of IS research methods from the pre-1989 period to the present. He argues that the publication of three paradigmatic articles in *Communications of the ACM* in 1989 is a turning point in the discourse on methods in our field, a break with the dominant paradigm and explicit recognition of the existence of multiple paradigms in IS research. Klein briefly sketches the intellectual journey of IS research from its beginnings a mere thirty years ago. Its Kuhnian revolution from the hegemony of the orthodox method of science to its apparent abandonment and the present blooming of a thousand flowers of interpretivism. He cautions, however, that "interpretivists of all flavors should heed the often forgotten dictum that data without theory is blind and theory without data is empty." His challenge to us for the next decade is "to inject more methodological controls into interpretivism so that its results can gain the trust of researchers and practitioners alike while not losing its flexibility and sensitivity to situational differences."

   Boland continues this critical reflection but takes us to an altogether different place. He examines how the notion of "space" has dominated, and still dominates, our field of research. He does a very interesting walkthrough of his own research to show how he attempted to make sense of images of "space." He recounts how he conceptualized the implementation problem as one in which mutual or shared understanding needs to develop. He shows how he attempted to make sense of the notion of "shared" understanding or knowledge through concepts such as frames and mappings. In the second section of the paper, Boland attempts to sketch out some of the reasons why the "spatialisation" of thought developed its prominence. He also introduces the work of Henri Bergson to show how the spatial triumphed over the temporal and what distortions this introduces. He concludes in a very frank way by stating: "So at this point I can only ask for giving more attention to the narrative mode without knowing exactly how to do so." He posits that the important issue for understanding implementation is to develop more temporal methods for representing and analyzing organizational phenomena. Methods that would "appreciate experience as it unfolds, that are sensitive to rhythm, tempo and construction in the flow of becoming."

As if in response to Klein's challenge, Baskerville and Lee take up the discourse on methodological controls in qualitative research. Specifically, they address the question of generalizing in qualitative research, commenting on the tentativeness of qualitative researchers to claim general contributions to the knowledge base of our field. In this piece, Baskerville and Lee outline different types of generalizing and discuss how each can lay claim to generality. Kaplan, Farzanfar and Freeman continue the critique of method with a discussion of the ethical dilemma the ethnographer faces in interviewing, recording and representing the worlds of the people he/she is studying. They present problems and findings from a study of the use of an intelligent interactive telephone system in a health care setting and discuss how individuals who use the service respond to it. Kaplan, Farzanfar and Freeman point to some open questions about attachment and alienation effects of IT and the ethical implications of some types of IT systems. Klein and Huynh address a fundamental problem of interpretive research, that is, how to systematically analyze and make sense of the mounds of empirical materials that are collected. Using empirical materials from a field study, they demonstrate how computer-aided language action analysis can assist in ethnographic analysis. They offer some advice on problems that researchers can expect when conduction such analysis.

2.2 Field Studies

Factors for successful implementation of ERP systems are outlined by Parr, Shanks and Darke. Hanseth and Braa discuss some of key issues of standardization of IT infrastructure that ERP systems, such as SAP, and collaboration systems, such as NOTES, impose on the implementing organization. Reporting on the first stage of a research program, Parr, Shanks and Darke seek to understand which factors might be necessary for the successful implementation of ERP systems. They identify ten candidate "necessary factors" for successful implementation, but admit that more research is needed to explore the relationship between these factors and broader contextual and processual issues. Hanseth and Braa provide us with a rich case study of Norsk Hydro, a very large manufacturing organization with diverse divisions and factories dispersed in many locations. The central "IS-Forum" and "Corporate Steering Group for IT" decide to establish a corporate-wide standard known as the "Hydro Bridge." The paper recounts a fascinating narrative of the attempts to make sense of what this means and how it will be, and is being, established. The narrative is essentially about the tension between our taken-for-granted notion of standards as universal objects shared by all in which there is no redundancy and no inconsistency, and the "always never *a* standard" of every attempt to make it work. Sauer, Johnston, Karim, Marosszeky and Yetton discuss opportunities and barriers to implementing collaboration technologies in the Australian construction industry. They argue that although inter-organizational collaborative technologies offer competitive benefits to the construction industry (especially from cost reduction by reengineering the supply chain), this industry is slow to adopt these technologies. They explain this phenomenon by identifying industry-level conditions, which have influenced the low level of IT-based collaboration, and suggest industry-level interventions, which could stimulate IT adoption in future.

In his paper, O'Donovan suggests organizational disposition as an alternative narrative to organizational culture for investigating organizational change. He draws on the work of Heidegger in an attempt to develop a rather subtle argument about the way in which individuals generate an intersubjective awareness, as a prevailing mood, of their current situation that renders the possibilities for action available even before they have articulated it. This mood provides the implicit logic for action, the organizational disposition. O'Donovan uses this notion to argue how implementer strategy, and user resistance, gets weaved together in ways that make some implementations succeed and other fail. O'Donovan examines how organizational disposition affects the implementation process in two case studies. Scheepers examines the roles that the various actors play in the implementation process. He identifies from the literature five key roles that one would expect to find in Intranet implementation projects, then proceeds to carefully analyze three case studies in very different contexts to examine the role players, and their importance, in the initiation and implementation of intranet technology projects. He concludes that there are indeed reasons to believe that many of the traditional role taxonomies do not apply in the new context. Sarker and Lee investigate the proposition that computerized BPR tools have a positive influence on business reengineering effectiveness. Conger and Schultze illustrate with a case study how genre theory can be used to develop a conceptual framework for studying the practice of electronic commerce. Karsten, Lyytinen, Hurskainen and Koskelainen use the concepts of boundary objects and perspective taking and perspective making to elucidate how individuals interact and solve problems via IT in highly specialized and distributed engineering activities. In their case study, they show how these concepts can be useful for understanding the issues that would need to be solved if a certain relatively disorganized document set were to be redesigned to be a part of a collaborative information system, connected to data in ERP and other formalized systems. In a fair amount of detail, they analyze how the "technical specification document" for a paper machine (which could be a 500 to 600 page document) functions as a boundary object for perspective taking and as a conscription device for perspective making. They conclude that, although their study gives no directions as to how this dilemma could be solved in practice, it does draw attention to the dialectic nature of boundary objects and conscription devices in perspective making and perspective taking. Spitler and Gallivan discuss the role of IT in supporting learning in a consulting organization. They use the concept of legitimate and peripheral participation to examine how a culture of IT competence emerges over time and how individuals might be empowered (or disempowered) by their IT competence.

An underlying theme of all these field studies is the extent to which IT systems intervene in the social fabric of organizations. Given this situation, one might expect implementation methodologies to be deeply sensitive to the social transformation that takes place when these technologies are implemented. No so, say Hemingway and Gough. Although debate for the next generation of development methodologies is ongoing, researchers still have problems incorporating the concept of social transformation in systems development methods and practice. Hemingway and Gough propose a general structure for such a methodology based on socio-cognitive theory and give an indication of its logic and possibilities. They acknowledge that their work is still very much "in progress" but defend it as a reasonable attempt to address not only theory but also development practice.

3. Concluding Remarks

While this volume can in no way sum up the contributions that IFIP W.G. 8.2 has made to the field of information systems during the last ten years, it does mark the end of the decade and millennium. It builds on a decade of achievements in research on IT and organizations and gives an idea of some of the unsolved problems that we will take into the next millennium. We hope that you find it informative and exciting to read. We hope that those attending this 1999 conference on New Information Technologies in Organizational Processes: Field Studies and Theoretical Reflections on the Future of Work find the discussions stimulating.

About the Authors

Ojelanki K. Ngwenyama is Associate Professor and Director of the Center of Excellence in ERP Research at Virginia Commonwealth University; Docent at University of Jyväskylä, Finland; Visiting Research Professor at Aalborg University, Denmark; and Extraordinary Professor at University of Pretoria, South Africa. He is a co-founder of the Collaboratory for Research on Electronic Work (CREW), University of Michigan. Between 1990 and 1997, Ojelanki was a faculty member of The University of Michigan Business School. He is a critical theorist whose research focuses on how people engage information technology in organizations. He holds a Ph.D. in Computer Science and Information Systems from The Thomas J. Watson School of Engineering, State University of New York. He is a member of the Editorial Board of the *Journal of the Association of Information Systems* and Associate Editor of the journal *Information Technology & People*. He is co-editor of the book *Transforming Organizations With Information Technology,* North Holland, 1994. Ojelanki can be reached by e-mail at ojelanki@isy.vcu.edu.

Lucas D. Introna is Lecturer in Information Systems at the London School of Economics and Political Science and Visiting Professor of Information Systems at the University of Pretoria, South Africa. He is associate editor of *Information Technology & People* and editor of *Ethics and Information Technology.* He is an active member of IFIP W.G. 8.2, The Society for Philosophy in Contemporary World (SPCW), the International Sociological Association WG01 on Sociocybernetics, and a number of other academic and professional societies. His most recent work includes a book, *Management, Information and Power*, published by Macmillan, and he has published widely in the best IS journals and conference proceeding on a variety of topics such as theories of information, information technology and ethics, autopoiesis and social systems, and virtual organizations. Lucas can be reached at l.introna@les.ac.uk.

Michael D. Myers is Associate Professor in the Department of Management Science and Information Systems at the University of Auckland, New Zealand. His research interests are in the area of information systems development, qualitative research methods in information systems, and the social and organizational aspects of information technology. His papers have appeared in a wide range of journals, including *Accounting, Management and Information Technologies, Communications of the ACM, Ethics and Behavior, Information Systems Journal, Information Technology and People, Journal of Information Technology, Journal of Management Information Systems, MIS Quarterly* and *MISQ Discovery.* He is co-author of two books, including *New Zealand Cases in Information Systems* (with J. Sheffield, Pagination Publishers, 2nd edition, 1992). Michael is Editor of the ISWorld Section on Qualitative Research, an Associate Editor of the *Information Systems Journal*, an Associate Editor of *MIS Quarterly*, and on the Editorial Boards of *Communications of the AIS, Information Technology & People* and *Journal of Systems and Technology.* Michael can be reached by e-mail at m.myers@auckland.ac.nz.

Part 1:

Critical Reflections

2 NATIONS, IDENTITIES, AND GLOBAL TECHNOLOGIES

Mark Poster
University of California, Irvine
U.S.A.

Abstract

As a political unit, the nation is facing an ever-expanding set of challenges. Modern systems of transportation and communication facilitate global exchanges of commodities, populations and information, often evading the borders and jurisdictions of the nation state. Faced with an increasingly interconnected globe, the nation may no longer be able to sustain its territorial hegemony. Nor may it be able to contain citizens within national identities. This talk examines the role of the Internet in these processes.

I examine how the Internet affects two features of the nation: sovereign borders and national identity. I contrast the role of print in the formation of national identity with that of communication in cyberspace. I surmise that in addition to a global economy, we may be seeing the emergence of a global culture with severely diminished prominence of the nation. I raise the question of a new global political unit and its possible relations to older forms of political community. Is the Internet a vehicle of U.S. domination or the basis of new political forms that combine the global and the local in new, possibly less hierarchical ways than in the past? What are the possibilities and challenges to the flow of information in this context?

Keywords: National identity, nation state, Internet, globalization, culture.

About the Author

Mark Poster is Professor of History and Director of the Critical Theory Institute at the University of California, Irvine. His most recent books are *The Information Subject* (1999), *Cultural History and Postmodernity* (1997), *The Second Media Age* (1995), and *The Mode of Information* (1990).

3 KNOWLEDGE AND METHODS IN IS RESEARCH: FROM BEGINNINGS TO THE FUTURE

Heinz K. Klein
State University of New York, Binghamton
U.S.A.

Abstract

The purpose of this paper is to identify some of the key challenges that the Information Systems research community needs to address in order to improve interpretive and critical research approaches in the next decade. In order to achieve this goal, a high level review of the principal trends in IS research methodology since the late 1960s set the stage for an analysis of challenges to be met in the next decade. The past trends in IS research methods are analyzed from the perspective of two questions: what is the implied concept of knowledge and what are the methods of inquiry preferred by various research methods schools? Even though the analysis attempted here should be viewed as tentative and preliminary, it leads to several provocative conclusions. The most significant of these concern the interrelationships between critical and interpretive research.

Keywords: Evolution of research methods in information systems, current research methods and their limitations, inquiry methods, ethical standards, rules in information systems development, ethnography, confessional genre, critical theory research, cross-paradigm approaches.

1. Introduction

When I first read the call for papers of this conference, I was struck by a paradox, which ultimately contributed to my acceptance of the invitation for a keynote speech to this conference. I am very grateful to the organizers for offering me this forum for articulating some of my recent reflections on the current state of IS (Information Systems) research. While my ideas are still very much in the making, their preliminary status will not keep me from offering some provocative propositions.

The call for papers of this conference refers to the common observation that the many IT innovations cause a wide range of social transformations. There is a note of alarm in bemoaning that "These radical social transformations of organizations are taking place át such a speed that they are overwhelming for academic researchers." What is needed are general concepts to classify the mind-taxing variations of organizational transformations into a parsimonious typology. Something like the table of elements would be ideal, because it also has predictive power. We also need abstract theories that allow reducing the complexity of organizational reality transformations into a few comprehensible patterns just as Newtonian axioms bring order into the diverse phenomena of mechanics.

Alas, the call for papers takes the reader in an entirely different direction, when it states that "we are interested in field studies that discuss social and organizational issues around the implementation and use of these new technologies in organizational processes." I find this paradoxical, because field studies are predominantly ideographic, i.e., they tend to report the differentiating details of cases and make only weak attempts at classification and generalization. If there are an "overwhelming" number of different social transformations, they will then be reflected in an equally overwhelming number of different field studies pursued with a bewildering variety of research method variations.

The hypothesis that I wish to explore with you in this paper is that the paradoxical turn, which I perceived in the call for papers of this conference, is by no means an accident. It has become typical for a great many conferences in industry and academia. Why is it that many scholarly institutions in IS currently appear to be shying away from the noble pursuit of theory? Is there widespread disenchantment, maybe even discontent, with theory and, if so, why? My purpose is to deal with these questions in a way that could support a communal reflection on the path that IS research has taken during the last three decades. Only by looking back at how we got here can we become more critically aware of the choices that we are about to make as we move into the fourth millennium. (I start my count with the first 1000 years before Christ, which brings the Presocratics and classical Greek philosophers into the elite of human knowledge creators.)

2. Approach

In reflecting upon three decades of IS research and the role that my research associates played in the evolution of the fundamental tenets underlying IS research, I am struck by how the meaning of the simple word knowledge has changed in a relatively short time for our discipline. Clearly one cannot consider the question "what do we mean by knowledge?" without also bringing the methods of knowledge creation into the bargain. Knowledge and the preferred methods of inquiry used to create knowledge form an inseparable unit. If one changes, so must the other. Methods of knowledge creation cover a wide variety of human behaviors, from the way families acquire knowledge for conducting their lives, from jury and court decisions to organizational knowledge creation and academic communities. For obvious reasons, I will by and large limit myself to discussing the kind of preferred methods of knowledge creation in the IS community that we typically call research methods. For a convenient collection of such research methods I refer to Mumford et al. (1985) and Nissen, Klein and Hirschheim (1991).

My approach will be to trace the interdependence of knowledge over the life of our discipline and relate the main stages of this evolution to broader philosophical ideas. I will then ask, what is the most recent turn in our search for knowledge both with regard to the characteristics that define what counts as knowledge and with regard to the ever-moving front of research methods? In particular, what do the most recent proposals for refining our research methods imply for the concept of knowledge? To clarify the basic plot of my investigation a little further, I would like to give an example of the inter-dependence between knowledge and research methods by referring to five historical figures. We all heard about some of the kingpins in the Western history of knowledge creation such as Plato, Euclid, Kant, Newton and Einstein. (Note that these figures come from two of the three millennia of knowledge creation in Western philosophy and science, one millennium B.C. and one after the year 1000. I did not dare to face the tricky question: who are the principal knowledge creators during the first 1000 years A.D.?) While all of these five historical figures contributed to our stock of knowledge and all still exert strong influence on modern thought, the methods they followed are vastly different and so is the concept of what they meant by knowledge. For example, Plato and Kant explicitly included ethics (the moral imperative) into the realm of knowledge, while the others did not. Moreover, the example makes it patently clear that research methods may change radically. The methods of the first four figures are no longer acceptable to their present-day successors.

The question I wish to pursue in some detail is who have been the kingpins of knowledge during the last three decades of IS research and what have been their preferred research methods? Ancillary questions are why do we, their successors, no longer follow these preferred methods (what are their shortcomings) and which methods have been substituted for them? And finally, where does this all take us as the third millennium of Western knowledge creation comes to an end?

How should we break down 30 years of IS research into a few major streams that can be discussed in the short space available for this paper? As far as the concept of knowledge is concerned, I intend to critically examine representative crown witnesses from the prevailing academic literature in IS with the dichotomous distinction between facts and norms. I shall inquire from the prevailing lines of research, whether they were primarily concerned with descriptive knowledge or whether their definition of knowledge was so broad as to include normative knowledge. The typical forms of normative knowledge are norms, rules and values. Rules are important in the practice of any profession—e.g., management, engineering, medicine, law—but they play a particularly prominent role in ISD (information systems development). To further characterize the kinds of rules that are typical for the normative side of knowledge, I shall apply Kant's distinction between rules skill, rules of prudence and categorical rules (to which I shall refer as ethical values). **Rules of skill** are concerned with the physical propensity and dexterity to carry out certain operations to achieve specified ends. **Rules of prudence** are concerned with judgments to achieve ends that informed and reasonable people would not question as being worthwhile, such as designing systems which are acceptable to their intended users. **Categorical rules** are concerned with choices where the ends themselves are in question, such as a choice between developing systems which are preferable to one set of stakeholders (e.g., workers on the shop floor) or another (e.g., a company's shareholders). Categorical rules are particularly important when our discipline has to deal with social value conflicts in practice and hence I shall refer to them as ultimate values or ethical standards. Klein and Hirschheim (1998) provide a fuller treatment of these ideas.

In order to analyze changes in preferred research methods over time, I shall draw on the seminal book by W. Churchman, *The Design of Inquiring Systems: Basic Concepts of Systems and Organization*, 1971. In this book, Churchman proposed that the principal methods of knowledge creation (inquiring method as he called them), can be broken down into five basic epistemological types: Cartesian, Lockean, Leibnizian, Kantian and Hegelian types of inquiry. Modern research methods are descendants of the last three combining and varying their characteristics. A key characteristic of modern research methods is that they all have abandoned the idea that knowledge begins with simple and clear entities, that we can start with basics that are obviously true and build more complex theory on secure foundations that are beyond doubt. This eliminates Cartesian and Lockean types of inquiry from serious consideration. Yet, early IS research, to some extent, did subscribe to Cartesian and Lockean inquiry ideals.

Of course, the five epistemological types of inquiry do not neatly follow each other in time, but are recognizable as literature pools and quoting cartels overlapping in time. However, a convenient landmark to break the last 30 years of IS research into two periods is the explicit recognition of multiple paradigms in widely read, so called premier journals. This occurred around 1989 because of two key events. One was the appearance of three "paradigmatic" articles in the *Communications of the ACM* in short succession: Banville and Landry, Hirschheim and Klein and, in opposition to these two, Denning et al. Of course, the recognition of multiple paradigms in the research establishment could not have occurred without some precursors such as Mumford et al. 1985, Chua 1986 and Galliers and Land 1987. These paved the way for a broader based attitude change in several research communities. There is a certain irony to temporal coincidence of the 1989 publications. Neither of the three sets of authors were aware of each other's projects even though the first two papers reached almost identical conclusions and there would have been plenty of time for contacts as the review process took several years. A second key event was the conversion of *MIS Quarterly* from a single paradigm journal to one that would restructure its review process so that contributions from alternative paradigms would receive a fair shake without jeopardizing high standards. Whereas there have always been "alternative community journals" (e.g., *Information, Technology and People,* edited by E. Wynn, has offered strong support for interpretivist research for quite some time), the journal culture began to shift toward multiple paradigm publication venues. Younger researchers often don't recognize that this amounted to radical, institutional change.

What follows is a review of my current state of thought on these issues. It is a first draft that cannot do justice to all the questions raised in the introduction.

The remainder of this paper previews some of the key ideas in draft form, which I shall develop more fully for the conference presentation. In the interest of brevity, I need to be highly selective, simplify many aspects and fall short of answering all the issues raised in the introduction. While I will draw on my knowledge of the historical evolution of IS research, I don't intend to write an authoritative history of IS research methodology, but merely highlight major epistemological shifts in IS research. An epistemological shift occurs if there are major changes in the concept of what counts as knowledge or the preferred research methods.

3. Archtypcial Patterns in the Pre-1989 Era of IS Research

The purpose of this section is to illustrate three observations about the general characteristics of IS research during the era from the mid-1960s to the end of the 1080s. First, while the MIS publications of the very early period dating back to mid-1960s were methodologically less rigorous, they exhibited much concern for both descriptive and normative knowledge with a variety of inquiry methods. A good collection of them can be found in Sanders (1970), but one of the best-known examples is probably Ackoff's 1967 classic paper on MIS-misinformation systems. The papers of this genre apply Lockean types of inquiry in that they rely on clear and simple observations and use common sense logic to establish plausible conclusions of a normative type. They often have the ring of stories; therefore, this stage in IS research has sometime been called the era of "great wise men." One of the surprising turns of recent research is the rehabilitation of the story concept as a legitimate research genre—but let me not get ahead of myself.

My second observation moves us further ahead into the 1970s when the so-called experimental school of IS research (beginning with the Minnesota Experiments, Dickson et al. 1977) substantially increased the requirements for methodological rigor, but at the same time the practical relevance of this type of MIS publications decreased. One of the most visible counter-reactions to this trend was the reorganization of *Communications of the ACM* a few years ago. Nevertheless, when judged by sheer numbers of paper titles, several studies concluded that 90% or more the body of knowledge published in journals falls into this category and it still is the dominant schools that attracts the most disciples. The principal mode of inquiry of this line of research is Kantian with almost exclusive emphasis on descriptive types of theoretical knowledge. It is Kantian, because this kind of research proceeded from formulating theoretical hypotheses to derivation of conclusions or prediction which could then be tested empirically using sophisticated statistical inference methods. Hence this kind of research, which is still very active, emphasizes an internal model with a complex logic that is used to make sense of outside data. It is commonly referred to as the positivist model of scientific research.

Parallel to the experimental methods of inquiry, several other lines of research flourished during the same era using research methods of a very different kind. Most of these were applied in the area known as ISD (information systems development). ISD research at that time created the knowledge core that defined the identity of our discipline as is apparent from the article by Ives, Hamilton and Davis (1980). (Do we still have such an identity defining knowledge core?) Very significant influences to improve ISD came from **action research** (e.g., Checkland 1972), which is still going strong. Action research equally emphasizes descriptive and normative knowledge at the level of rules of skill and rules of prudence in our field, but generally it stops short of contributing critical evaluation of ethical standards. In the early years, a substantial amount of research on ISD centered on the **participation debate** in ISD (Mulder 1971; Mumford 1983, 1984; Greenbaum and Kyng 1991). In this context, important intellectual and political insights were contributed to our understanding of the broader social issues surrounding ISD by the so-called **trade union or collective resource school** within the **Scandinavian School.** (Kyng and Mathiassen [1982], Kyng [1991], Bansler [1989], Bjerknes and Bratteteig [1995], and Mathiassen [1999] present good literature analyses of major contributions from the Scandinavian School to IS research.)

Another major impetus for advancing the state of the art in ISD came from engineering research into the development of the first generation **computer-based tools of ISD** (cf. Couger, Coulter and Knapp 1982 for a review of several projects in this area). All of these streams research, which often were carried out separately without the benefit of mutual communication and debate, contributed to a larger body of knowledge which became known as **alternative methods and tools of ISD**. The **prototyping and decision support systems research** (e.g., Keen and Scott-Morton 1978; Budde et al. 1984) is also part and parcel of the extensive research on ISD before 1990.

The core knowledge created by all these research streams incorporates both descriptive accounts of how ISD is actually carried out in practice, empirically tested hypothesis among environment, process and ISD outcome variables (Ives, Hamilton and Davis 1980) and normative knowledge including ethical standards. Figure 1 gives a summary of some of the principal value standards that emerged from the debate on the ultimate goals of ISD. The most widely espoused value standard is, of course, the private enterprise ideal. It insists that the common good is best served if IS improves the competitiveness and profitability of private enterprise and is subject to some minimal legal constraints on privacy, job security, etc. All else is secondary. The research methods of the research on ISD defy a simple description. It spans several paradigms (cf. Hirschheim, Klein and Lyytinen 1995 for more extensive discussion) combining elements of Kantian, Leibnizian and Hegelian, i.e., dialectical, methods of inquiry.

The best examples for Leibnizian inquiry are found in the literature from the computer simulation movement. It built very complex models and investigated the consequences of the complex assumptions that were built into the model. Hence it started with complex assumptions the truth of which was unclear ("contingent truths"). By running the model and looking at its output, it was hoped that the model could eventually be improved. This idea is basically Leibnizian. As we all know, the enthusiasm for computer simulation as a research method died in the early 1980s.

Third, in the 1980s, the narrow epistemological assumptions of the positivist line of IS research became ever more apparent to researchers who were aware of the philosophical critiques of positivism (cf. the literature analysis in Klein and Lyytinen 1985). Eventually this led to a valiant attack on positivism from several directions. The paradigm controversy in the IS literature made many more researchers aware of the importance of dialectical research methods and established the notion of methodological pluralism, which characterized the debate in the 1990s.

(A) **Socio-technical Ideal.** The principle objective is to optimize the interrelationships between the social and human aspects of the organization and the technology used to achieve organizational goals. Both the quality of working life (satisfaction of human needs at work) and profitability or system efficiency are to be improved through a high degree of fit between job characteristics (as defined by work design) and a limited set of human needs (as defined by social and psychosomatic profiles).[1]

(B) **Decision Support Systems Ideal**. The final criterion is "to help improve the effectiveness and productivity of managers and professionals." A necessary condition for achieving this is to gain user acceptance and actual systems use. In some parts of the literature, this is seen as a worthwhile goal in itself. In any case, the emphasis is on tailoring the system design to the user's needs and preferences. Sophisticated techniques for analysis of personality traits or other psychometric variables such as cognitive style, locus of control, motivational type, attitudes, etc., are proposed as design tools. Thorough "understanding" of the user's problems and a strategy of mutual trust building through highly participative systems development in a series of small, adaptive steps with rapid feedback through demonstrations are seen as the "right" approach.

(C) **Dialectical Inquiry Ideal**. Above all, information systems must be designed such that they produce maximal motivation to generate "objective" knowledge through the synthesis of the most opposing points of view, each supported by the best available evidence. Truth and objectivity are of prime importance and can only be found through an adversary system in which the competing points of view are confronted such as in a court of law. Peace, willingness to compromise, or consensus creates dangerous illusions, which threaten the objectivity of knowledge and justice (cf. Churchman 1970, 1981; Mason 1969; Cosier et al. 1978).

(D) **Participatory Design Ideal**. Emphasizes that the process by which systems are developed may be more important than the features of the final product, because "people should be able to determine their own destinies." Hence the ultimate moral value to be achieved through participation is human freedom, which then leads to such other goods such as motivation to learn, genuine human relationships based on equality and commitment to what one has freely chosen to accomplish (Mumford 1983; Land and Hirschheim 1983).

[1]Proponents of the socio-technical ideal usually stress that it will improve profitability through higher worker satisfaction (and ignore a possible conflict between the two). The dialectical inquiry (DI) ideal has been tried in the context of improving private enterprise planning (Mason 1969) and the decision support system (DSS) approach stresses "managerial effectiveness" over technical cost savings. DI and DSS could be treated as end goals, but are primarily seen as instrumental for achieving other final ends. The socio-technical and private enterprise ideals are examined in detail in Klein (1981).

Figure 1. Ultimate Goals or Ethical Standards Identified in the Literature on Methods and Tools of ISD (Klein 1981)

4. Beyond 1989: Dialectics and Methodological Pluralism

Even though Mason introduced the basic ideas of dialectical inquiry (originally due to Hegel) into the management literature as early as 1969 in *Management Science*, it took about two decades for dialectical concepts to exert major influence on IS research. Dialectical inquiry insists that the growth of knowledge benefits most from social processes that are propelled by conflict and passion. A dialectical inquiring system consists of at least two actors with differing worldviews or perspectives by which they interpret the world in conflicting ways. Each actor has access to a separate data pool (her accepted "facts") that only partly overlap with the data of the other actor or actors. Each actor possesses her own logic and modes of reasoning, again not completely congruent. Knowledge emerges from debating the contradictions among the beliefs, inference methods and data of each actor. By eliminating some of the contradictions in reformulating the worldviews of some of the actors, a higher level of knowledge is reached. This new worldview incorporates the best of its antecedents and is called a "synthesis." The synthesis is the starting point for a new cycle of debate with an emerging opposing viewpoint and the process repeats.

The Burrel and Morgan (1979) classification of research paradigms in sociology provided a convenient description of prevailing worldviews among researchers in the 1980s. Hence, when the aforementioned papers by Banville and Landry and by Hirschheim and Klein appeared, they offered a convenient intellectual rationalization for what many IS researchers intuitively had come to accept: that there are rather different approaches to ISD and other issues in IS research. By contrast, the third paper (Denning et al.), formulating the opposing positivist viewpoint as a research paradigm for Computer Science, was almost completely ignored in IS. This further widened the intellectual gap with Computer Science. Instead of debate about paradigms, a quiet shift to pluralism occurred. The idea that ruled the first 20 years of IS research, namely that the one and only orthodox method science fits all, was abandoned within the relatively short space of a few years—much to the surprise of this author.

The result of this was the emergence of several alternative research streams as viable Ph.D. dissertation projects and "tenurable" research programs. It has now become accepted that there are three research paradigms in IS research, namely the positivist, the interpretivist and the critical. The field is now also willing to consider such radical alternatives as post-modernism or deconstructionism, at least tentatively. Nevertheless, when counting sheer numbers of disciples in each paradigm, some version of positivism is still the most frequent type of research. However, publications that are guided by interpretivist modes of inquiry are on the rise. There is a dearth of critical inquiries. Most critical research published is concerned with fundamentals, such as Klein and Lyytinen (1991), Hirschheim and Klein (1994) or Ngwenyama and Lee (1997). It is my contention that an analysis of the reasons for this grants little comfort and provides future researchers with major challenges. Nevertheless, many researchers, including positivists nowadays will agree that if interpretivist and critical research publications are weighed in as a new body of knowledge that has emerged in the 1990s, as a whole it successfully challenges the knowledge and research methods monopoly of positivism. Even such a bastion of positivism in our field as the Decision Sciences Institute sponsored a special round table panel session in March 1999 (North Eastern Decision Sciences Institute, New Providence) with the provocative title: "Organizational

Research in the 21st Century: Have We Made a Qualitative Shift?" Many of us can no longer avoid the impression that we are currently living through a "scientific revolution" in our field in the sense of Kuhn (1970). But where do we go from here?

5. Stories Lost, Stories Found: Challenges for the Next Decade

Judging by the recent surge of interpretivist dissertations, conferences and journal publications, the ethnographic and intensive field study species of interpretivists are enjoying the equivalent of a communal honeymoon. However, there are several points that I would like to make in the conclusion, which give us little reason for comfort. First, the field started with story telling that was then debunked as "unscientific." It now appears has if IS research has come full circle by returning to new forms of story telling (the politically correct term is, of course, "narratives'). Granted that the new story telling movement can point to much better and more explicit philosophical grounding than "the great wise men" had for their stories, I still wonder to what extent all the recent converts truly understand the intellectual foundations of their newly found faith. If one asks, just what are the methods by which the new breed of storytellers distinguishes themselves from their ancestors, one realizes that the debate on proper methods for interpretivist research has barely begun (e.g., Klein and Myers 1999; Schultze 1999). If and when we get such a debate off the ground with broad participation, it offers the chance of rapid, dialectical growth of knowledge. However, along with this chance for a debate also looms the danger of sectarianism—the split in subcommunties, which prefer to ignore each other rather than to engage each other in critical dialogue.

My second conclusion concerns the reasons why there appears to be a surge of interpretivist research as is so evident from this and the preceding Philadelphia conference (see Lee, Liebenau and DeGross 1997). While there is room for many different types of answers, I offer the following conjecture. To a major part, these reasons have to do with the failure of positivist research to offer a few broad theories that offer general orientation and bring some measure of order to the perpetual confusion in our field. Alas, interpretivists have little reason to be gleeful about this. They are even weaker in theory formulation than the positivists. Should it turn out that they are equally unable to deliver the goods, they may fall into disrepute much quicker than the attraction of positivism is waning. It may turn out that neither emperor has any clothes.

Third, what about critical theory research? While it has strengths exactly in the areas in which interpretivism is weak, i.e., theory formulation and normative knowledge, its empirical grounding is extremely weak. In fact, one can only be astonished how an eminent philosopher like Habermas deals with this issue. Moreover, if one asks which methods can be taught to aspiring critical researchers, one draws almost a blank card. There appears to be no research methods literature on critical research. This is evident from Forester (1992). It will take a whole generation of researchers to create such a literature. Some beginnings can be found in Klein and Truex (1996) and Klein and Huynh (1999, in this proceedings), but the amount of work will need to increase dramatically from many different quarters if further progress is to be made; yet the successful outcome is uncertain no matter how many join this effort.

Fourth, what about a possible union of critical research and interpretivism? In this area, I see great research opportunities, but let me first offer a few words of caution. The historical and conceptual connections between various forms of critical theory interpretivism are complex and therefore not well understood, maybe even by professional philosophers. I am therefore very skeptical if current attempts to integrate the two are founded on a clear understanding of their intrinsic connections. Why then do researchers interested in critical theory reach out to linguistic theory on the one hand (as is evident from my own publications) and to interpretivist field studies on the other (cf. Ngwenyama and Lee 1997)? It seems to me that it is merely the lack of a recognized stock of critical methods that provides the primary motivation for this. If so, current attempts to integrate these different conceptual bases are merely liaisons of convenience, because they are not based on an explicit reconstruction of the conceptual foundations that could give the newly found relationships staying power. Without proper theoretical foundations, it is merely of matter of convenience, if not desperation, to grab whatever avenue offers itself for data collection to experientially ground critical or other interpretivist ideas. Any port in a storm! It is convenient, because interpretivist methods of data collection and analysis provide a relatively safe haven, appeasing the researchers' understandable human need for some predictability and order in their work. However, unless the union can be based on a reasoned understanding that interpretivist and critical assumptions are at least partially compatible, the potential intellectual incongruence between the interpretivist approaches to data collection and the critical theory base could become the Trojan horse, which brings down the whole integration project.

Hence my fifth point is to call for serious work to construct the necessary theoretical and methodological bridges between critical theory and philosophy of language and interpretivism. This is an immense challenge for the IS research community. Clearly, critical theory is much more theory oriented than interpretivism, because it brings with it a strong legacy of Kantian a priori, at least in the version of Habermas' Critical Social Theory. (I offer this claim without substantiating it here.) Moreover, with its emphasis on a communicative orientation (or interest in human understanding), it explicitly relates to hermeneutics, which is also the heart of interpretivism. Therefore, I consider the full development of all the potential relationships between interpretivism and critical theory as one of the most fruitful avenues for future research. This union might very well be poised for a honeymoon, because of conceptual interrelationships in the underlying intellectual foundations of both critical social theory and interpretivism. For example, both relate to the philosophy of language (albeit in different ways, notice the use of speech act theory by Habermas), hermeneutics (e.g., Gadamer) and lifeworld philosophy.

However, none of the above addresses the shortcomings of interpretivism to contribute to the growth of normative knowledge. Its reluctance to address norm and value issues appears to make interpretivism incompatible with critical theory. I hope that future conceptual developments will somehow be able to overcome this limiting characteristic of interpretivism. For if not, I am sure of the following. IS theory will remain chronically incomplete if we as researchers simply ignore the immense importance of normative knowledge for practice and the general betterment of the conditions of human existence.

In summary, the first challenge for the next 10 years is to inject more methodologi-cal controls into interpretivism so that its result can gain the trust of researchers and practitioners alike while not losing its flexibility and sensitivity to situational differences. The second challenge for interpretivist research is to engage in more theorizing to provide generalizations, which promise to yield some order and predictability for our field. The third challenge is to find variations of the interpretivist approach, which allow entering a strong theoretical union with critical and other social theories. The hope is that interpretivist research can help with the experiential grounding of theoretical constructions and social theories of various *couleurs* could provide the theoretical bases to inspire more generalization and abstraction in interpretivist studies. Hopefully this enterprise will not get stuck in "middle range hypotheses" like positivism. In order to yield broad theoretical orientations, interpretivism must not follow positivism into the swamp of endless trivial hypothesis proliferation. The last and maybe biggest challenge for the future of IS research is to come to grips with the need for normative knowledge in our field that goes beyond rules of skill.

I know that some readers will react to these ideas with "here we go with yet another meta-narrative." My reply is that I do not seek the great, infallible truth, but merely some fallible theories in the sense of Popper's "Conjectures and Refutations" (1963), but of sufficiently broad scope. Interpretivists of all flavors should heed the often forgotten dictum that data without theory is blind and theory without data is empty. At the moment, interpretivists are more at risk to cause blindness than emptiness and the reverse is true for critical theorists.

References

Ackoff, R. "Management Misinformation Systems," *Management Science*, December 1967, pp. 147-156.

Banville, C., and Landry, M. "Can the Field of MIS be Disciplined?" *Communications of the ACM* (32:1), 1989, pp. 48-60.

Bansler, J. "Systems Development Research in Scandinavia: Three Theoretical Schools." *Scandinavian Journal of Information Systems* (1), 1989, pp. 3-20.

Bjerknes, G., and T. Bratteteig, T. "User Participation and Democracy: A Discussion of Scandinavian Research on System Development," *Scandinavian Journal of Information Systems* (7:1), 1, April 1995, pp. 73-98

Budde, R.; Kuhlenkamp, K.; Mathiassen, L.; Zullighoven, H. (eds.). *Approaches to Prototyping.* Berlin: Springer-Verlag, 1984.

Burrell, G., and Morgan, G. *Sociological Paradigms and Organizational Analysis.* London: Heinemann, 1979.

Chua, W. F. "Radical Developments in Accounting Thought," *The Accounting Review* (61), 1986, pp. 601-632.

Churchman, C. W. *Challenge to Reason.* New York: Delta Books, 1970.

Churchman, C. W. *The Design of Inquiring Systems: Basic Concepts of Systems and Organization.* New York: Basic Books, 1971.

Churchman, C. W. *The Design of Inquiry Systems.* New York: Basic Books, 1981.

Cosier, R. A; Ruble, T. L.; Aplin J. C. "An Evaluation of the Effectiveness of Dialectic Enquiry Systems," *Management Science* (24:14), October 1978, pp. 1483-1490.

Couger, J. D.; Coulter, M. A.; and Knapp, R. W. *Advanced System Development/Feasibility Techniques.* New York: Wiley, 1982.

Denning, P.; Comer, D.; Gries, D.; Mulder, M.; Tucker, A.; Turner, A. J.; and Young, P. "Computing as a Discipline: Final Report of the Task Force on the Core of Computer Science," *Communications of the ACM* (32:1), January 1989, 1, pp. 9-23.

Dickson, G. W.; Senn, J. A.; and Chervany, N. L. "Research in Management Information Systems: The Minnesota Experiments," *Management Science* (23:9), May 1977, pp. 913-923.

Forester, J. "Critical Ethnography: On Fieldwork in a Habermasian Way," in *Critical Management Studies*, M. Alvesson and H. Willmott (eds.). London: Sage Publications, 1992, pp. 46-65.

Galliers, R. D., and Land, F. L. "Choosing Appropriate Information Systems Research Methodologies." *Communications of the ACM* (30:11), 1987, pp. 900-902.

Greenbaum, J., and Kyng, M. (eds.). *Design at Work: Cooperative Design of Computer Systems.* Hillsdale, NJ: Lawrence Erlbaum Associates, 1991.

Hirschheim, R., and Klein, H. "Realizing Emancipatory Principles in Information Systems Development: The Case for ETHICS," *MIS Quarterly* (18:1), March 1994.

Hirschheim, R., and Klein H. K. "Four Paradigms of Information Systems Development," *Communications of the ACM* (32:10), 1989, pp. 1199-1216.

Hirschheim, R.; Klein, H. K.; and Lyytinen, K. *Information Systems Development and Data Modeling: Conceptual and Philosophical Foundations.* Cambridge, England: Cambridge University Press, 1995.

Ives, B.; Hamilton, J. S.; and Davis, G. B. "A Framework for Research in Computer-based MIS," *Management Science*, September 1980, pp. 910-934.

Kant, I. *Groundwork of the Metaphysics of Morals*, translated and analyzed by H. J. Patton. New York: Harper Torch Books, 1964.

Keen, P., and Scott-Morton M. S. *Decision Support Systems: An Organizational Perspective.* Reading, MA: Addison-Wesley, 1978.

Klein, H. K. "The Reconstruction of Design Ideals," paper presented at The Institute of Management Sciences Conference, Toronto, May 1981.

Klein, H. K., and Hirschheim, R. "The Rationality of Value Choices in Information Systems Development," *Foundations of Information Systems* [electronic journal: http://www.cba.uh.edu/~parks/fis/fis.htm], http://www.cba.uh.edu/~parks/fis/kantpap.htm, 1998.

Klein, H. K., and Huynh, M. Q. "The Potential of the Language Action Perspective in Ethnographic Analysis," in *New Information Technologies in Organizational Processes: Field Studies and Theoretical Reflections on the Future of Work*, O. Ngwenyama, L. Introna, M. Myers, and J. I. DeGross (eds.). Norwell, MA: Kluwer, 1999.

Klein, H. K., and Lyytinen, K. "The Poverty of Scientism in Information Systems," in *Research Methods in Information Systems*, E. Mumford, R. Hirschheim, G. Fitzgerald and A. T.Wood-Harper (eds.). Amsterdam: North-Holland, 1985.

Klein, H. K., and Lyytinen, K. "Data Modeling: Four Meta-Theoretical Assumptions," in *Software Development and Reality Construction*, C. Floyd, R. Budde and H. Zuellinghoven (eds.). Berlin: Springer, 1991, pp. 203-219.

Klein, H. K., and Myers, M. "A Set of Principles for Conducting and Evaluating Interpretive Field Studies in Information Systems," *MIS Quarterly* (23:1), March 1999, pp. 67-93. *Quarterly*, 1999.

Klein, H. K., and Truex, D. P. "Discourse Analysis: An Approach to the Analysis of Organizational Emergence," in *The Semiotics of the Work Place*, B. Holmqvist, B. P. Andersen, H. K. Klein, and R. Posner (eds.). Berlin: W. DeGruytner, 1996, pp. 227-268.

Kuhn, T. S. *The Structure of Scientific Revolutions.* Chicago: University of Chicago Press, 1970.

Kyng, M. "Designing for Cooperation: Cooperating in Design," *Communications of the ACM* (34:12), 1991, pp. 65-73.

Kyng, M., and Mathiassen, L. "System Development and Trade Union Activities," in *Information Society for Richer for Poorer*, N. Bjørn-Andersen (ed.). Amsterdam: North-Holland, 1982.

Land, F., and Hirschheim, R. "Participative Systems Design: Rationale, Tools and Techniques," *Journal of Applied Systems Analysis* (10), 1983.

Lee, A. S.; Liebenau, J.; and DeGross, J. I. (eds.). *Information Systems and Qualitative Research.* London: Chapman & Hall, 1997.

Mason, R. "A Dialectical Approach to Strategic Planning," *Management Science* (15), April 1969, pp. B403-B414

Mathiassen, L. "Reflective Systems Development," accepted for publication in *Scandinavian Journal of Information Systems,* 1999/2000.

Mulder, M. "Power Equalization Through Participation?" *Administrative Science Quarterly* (16:1), 1971, pp. 31-38.

Mumford, E. *Designing Participatively.* Manchester, England: Manchester Business School, 1983.

Mumford, E. "Participation—From Aristotle to Today," on *Beyond Productivity: Information Systems Development for Organizational Effectiveness*, T. Bemelmans (ed.). Amsterdam: North Holland, 1984, pp 95-104.

Mumford, E.; Hirschheim, R.; Fitzgerald G.; and Wood-Harper, A. T. *Research Methods in Information Systems.* Amsterdam: North-Holland, 1985.

Ngwenyama, O. K., and Lee, A. S. "Communication Richness in Electronic Mail: Critical Social Theory and the Contextuality of Meaning," *MIS Quarterly* (21:2), June 1997, pp. 145-167.

Nissen, H. E.; Klein, H. K.; and Hirschheim, R. (eds.). *Information Systems Research: Contemporary Approaches and Emergent Traditions.* Amsterdam: North Holland, 1991.

Popper, K. *Conjectures and Refutations.* London: Routledge & Kegan Paul, 1963.

Sanders, D. H. (ed.). *Computers and Management.* New York: McGraw Hill, 1970.

Schultze, U. "A Confessional Account of an Ethnography About Knowledge Work," accepted for publication *MIS Quarterly*, 1999.

About the Author

Heinz K. Klein earned his Ph.D. at the University of Munich and was awarded an honorary doctorate by the University of Oulu, Finland, for his academic contributions to the IS faculty's research program. He is currently Associate Professor of Information Systems at the State University of New York, Binghamton. Well known for his contributions to the philosophy of IS Research, foundations of IS theory and methodologies of information systems development, he has written articles on rationality concepts in ISD, the emancipatory ideal in ISD, principles of interpretive field research, alternative approaches to information systems development and their intellectual underpinnings. His articles have been published in the best journals of the field such as *Communications of the ACM, MIS Quarterly, Information Systems Research, Information Technology and People, Decision Sciences*, and others. He has also co-authored or edited several research monographs and conference proceedings in IS. He serves on the editorial boards of the *Information Systems Journal, Information, Technology and People* and the Wiley Series in Information Systems. He can be reached at hkklein@binghamton.edu.

4 THE TYRANNY OF SPACE IN ORGANIZATIONAL ANALYSIS

Richard J. Boland, Jr.
Case Western Reserve University
U.S.A.

Abstract

We want to understand organizational process as the temporal making of meanings, but our vocabulary for doing so is predominantly spatial. Some mistakes this has led to in my own research are reviewed, and the hope for a more thoroughly temporal mode of analysis based on an actor' ongoing narrativization of experience is explored.

Keywords: Schema, frame, narrative, spacialization, temporal.

> "We all secretly venerate the ideal of a language which in the last analysis would deliver us from language by delivering us to things."
> (Merleau-Ponty, *The Prose of the World*,1973, p. 4)

1. Introduction

Our everyday, metaphorical use of space when discussing cognition seems to be harmless enough—helpful even. We use space as a metaphor in our thinking about thinking all the time, especially the metaphor of thinking as a journey in an enclosed space and the metaphor of an argument as a container (Lakoff and Johnson 1980). We speak, for instance, of our line of thought, our areas of interest, our field of study, or our domain of expertise. We also speak of thinking as taking place within problem spaces (Simon 1979,1996). Ideas are then objects in a problem space. We hold ideas, we entertain ideas, we examine ideas, we compare ideas, and we turn them over in our heads. In order for us to do this with ideas, of course, they must be available to us in our

thinking in a very special way. They must be located or locatable within a space in order
for us to be handling them, observing them, laying them side by side for comparison, or
exchanging them in these ways. Taking ideas to be objects located in space and thinking
of movement in a problem space is part of a larger process of using spatial imagery to
understand our temporal experience as conscious agents (Bergson 1911, 1920).

In this paper, I discuss some implications of the familiar and everyday ways in
which we spatialize our discussion of thinking and ideas—of the way we use spatial
imagery to "get a handle on" the thinking of managers—in order to point out some of
the undesirable consequences which I believe are not so familiar. This paper began with
a series of uneasy feelings that developed over many years in doing my research on
information use and information design in organizations. For a number of years, I was
doing research that kept "backfiring" because the images of space that were implicitly
guiding it didn't seem to yield the expected results. Lately, I have begun to find a
vocabulary for discussing this sense of unease as having to do with the almost exclusive
use of spatial metaphors and the need for more temporal and processual ones. I have
also found a possible antidote for the excessive reliance on spacial imagery and the way
it encourages an understanding of cognition as information processing in Jerome
Bruner's (1986, 1990) more temporally sensitive work on narrative modes of cognition.

I will present my argument by recounting the process I myself went through in
trying to make sense of information design and use in organizations. In overview, I will
argue that our use of terms such as frames, schemas and paradigms is in many ways
misleading. They all are derived from the practice of spatializing our thinking about
thinking. They suggest the existence of bounded spaces that reside in the minds of
organization members which contain identifiable constructs and their interrelations.
They lead us to think of problems in organizations as often being created by an absence
of sufficient "shared knowledge," "shared understanding" or "shared meaning." They
lead us to propose that the creation of such "shared" cognitions will provide a solution
to many organizational problems.   Phrases such as "shared understanding" are
particularly potent examples of the way we spatialize cognition. I assume that in order
for understandings to be shared, we must first allow that an individual's cognition is
composed of discrete concepts or idea units and that they are located in a cognitive space
that is homogeneous or sufficiently compatible with the cognitive space of other
individuals such that they can be compared and found to be matching. I see no reason
to take this seriously, yet we continue to see in organization studies that success or
failure in organizations is attributed to the presence or absence of "shared meaning."

2. Some Studies of Frames, Framing, Reframing, and Frame Shifting

I will organize my argument by telling the story of my own journey through the world
of organizational research. It is a tale of lost beginnings, a tale of studies conducted to
search for answers to questions that were in the end called into question themselves. It
all began in 1973 with an interest in the "problem of implementation" in operations
research and management science. Very clever and seemingly helpful decision making
tools were being developed in response to management requests for assistance from
operations research and management science, but managers weren't using them.

Similarly, vast sums were being spent on the development of computer-based information systems, again, at the request of management, but in the end these systems were failing to be completed or used as intended. One popular way of posing this problem was provided by Churchman and Schainblatt (1965), who described it as a problem of failing to achieve "mutual understanding" between computer system analysts and the intended users of these systems. I began working on this "problem of implementation" and the search for mutual understanding by drawing on some notions from organization development and trying to create patterns of interaction between analysts and users that would tend to develop systems better attuned to management needs.

In that first study (Boland 1978), I began with the idea that system users and system analysts had different schemas or frames and the problem of implementation had to do with one frame dominating over the other (Argyris 1971), thus reflecting a lack of mutual understanding in the resulting system design. I focused on a key step in system design: the identification of information requirements, or the data that the new system was expected to produce either on a recurring basis in the form of standard reports, or on an ad hoc basis in response to a manager's special request. I characterized the traditional technique of system development as being led by the system analyst, and documented the best practices for interacting with users that were suggested by system analysis textbooks. These practices were then formalized as a set of protocols to be used by the analyst for interacting with the user. These protocols included guides for asking good questions, analyzing data, and developing a model of the user, the organization and their information requirements. I anticipated this would take place within the schema or frame of the analyst.

As an alternative to this, I used some techniques loosely adapted from organization development to create a set of protocols for interaction that would induce a team approach in which both the analysts and the users schemas or frames were involved equally in determining the information requirements. In this protocol, no one asked questions—instead, they both revealed things about their values, ideals and experiences to each other, then jumped to conclusions about what the computer system should be like, then critiqued each others' suggestions. I tested the two sets of protocols by having system analysts from industry and nurses from surgical floors of a hospital work in pairs to define the nurses' information requirements.

I anticipated, among other things, that the alternative protocols would result in greater mutual understanding, with each team member being better able to adopt the schema or frame of the other. As a result, I expected that the teams using the alternative protocols would develop better information requirements, as judged by a panel of nursing experts. The results confirmed these expectations. But in analyzing the designs that the different analysts had proposed, I was struck by the realization that they had all, in a sense, done equally well in solving the design problem that they had posed for themselves, but they had posed different problems to solve. Also, the panel of nursing experts was just one possible way of scoring these results. Other judges, such as finance directors or doctors, might have scored them much differently. What was it that made the nurses' frame the correct one for scoring these ideas? The analysts showed a variety of ways of framing the nurses' problem, each one led them to a different statement of the problem they were to design a solution for. This realization awoke in me an awareness that what had been going on was not so much a problem solving exercise as a problem framing exercise. The protocols didn't so much affect the ability to solve the nurses' problem as it affected the way the nurses' situation was turned into a problem. So it was not the problem solving performance that was being mediated by these different interaction protocols so much as it was the problem framing performance.

This led me to an exploration of the social construction of reality (Berger and Luckman 1967), symbolic interactionism (Blumer 1969), anthropological studies of symbolism (Sperber 1974), and a concern with biases from attribution theory (Kelley 1973) in which observers see others as facing situations that were more understandable, more certain and more subject to their control than the actors themselves did in those situations (Boland 1979). At the time, I was at the University of Illinois at Urbana-Champaign and privileged to be working with Lou Pondy. He and I did two studies together in which we explored problem framing in the context of accounting and budgeting. In the first study (Boland and Pondy 1983), we used Scott's (1981) distinction between "rational" and "natural" system models of organizing to explore budget cutting in two case studies. Rational system models "assume managements are confronted with an objectively knowable, empirically verifiable reality that presents demands for action" (Boland and Pondy 1983, p. 223). Using these models, managers assess cause effect relations and take action based on the demands of the situation. Natural system models, in contrast,

> see managements as responsible agents who interact symbolically and, in so doing, create their social reality and give meaning to their ongoing stream of experience. Problems are not simply presented to managements, problems are constructed by them. [Boland and Pondy 1983, p. 223]

Lou coined the phrase "Genuine Union" for the way we saw both the rational and natural models being used by managers in an alternating sequence of figure/ground relations. We were eager to argue against contingency theory and propose that both models were always in use, one always providing the background as the other was brought to center stage. It was not a contingent either-or but a genuine union both-and. In the budget cutting meetings we studied, we read managers dialogue as being within one model for a while, but then switching to the other and reversing their figure/ground relation.

In the second study, we analyzed the transcripts from a budget cutting meeting in a public school district that Lou had been studying (Boland and Pondy 1986). In making our interpretation of those transcripts, we used a coding system in which we recognized distinct modes and models of cognition and coded the dialogue based on the mode-model combination that was evident. For the modes, we used "instrumental" and "symbolic" modes of cognition (derived from Scott's rational and natural models), and we added the four decision models (fiscal, clinical, strategic and political) identified by Meyer (1984) in his study of hospital decision processes. Once again, we identified a series of switches among mode/model combinations, as shown in Figure 1 for the discussion of a budget line item. This strengthened our belief in the idea of a "Genuine Union" between instrumental and symbolic modes of thought, and of the importance of figure/ground reversals between them during decision dialogue. We also came to see the importance of shifting among decision models as well. The decision dialogue seemed to be taking place through a series of conversational transitions, so that knowledge

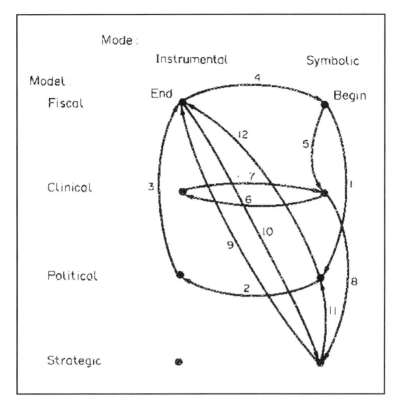

Figure 1. Map of Topic 1, "Athletic Budget," 561 Lines

creation was a journey moved forward by shifting among mode-model combinations through a series of conversational transitions. Participants seemed to carry meaning making through the transitions, and experience the discussion as continuous in the face of its continuous transition.

So far, the work I have reported had been a kind of strange dancing with frames. It started with a naive sense that different frames were held by different kinds of experts and the hope that different patterns of interaction could somehow meld them together as opposed to letting one of them dominate. It developed into an awareness of problem framing and the symbolic action of individuals in giving meaning to situations, as they actively frame them. From there, it evolved into an appreciation of the multiple frames available to actors and the process of working within frames versus working between frames. From there, we added the process of dynamically shifting among frames during conversation—with a figure/ground image of this shifting—as being important for understanding how group decision processes work. But all the time, the notion of a frame or a schema was there as an assumption. Whether the actor is assumed to be within a frame, in the act of framing, or shifting among frames, the idea of frame was still there. In retrospect, I am reminded of the exasperation revealed in *Philosophical*

Investigations when Wittgenstein looked back on his "picture theory" ideas from the *Tractatus*:

> 103...Where does this idea come from? It is like a pair of glasses on
> our nose through which we see whatever we look at. It never occurs
> to us to take them off.

Shortly after this work with Lou Pondy, I continued exploring frames with Ralph Greenberg, another colleague at Illinois. We looked at the role that metaphors might play in the problem framing process by inducing subjects to interpret a situation using either a mechanistic or an organic metaphor (Boland and Greenberg 1992). We induced the metaphors by first getting groups of subjects to identify examples either of machines (mechanistic) or of plants or animals (organic) using a nominal group technique to generate a list of examples. Then, we asked them to identify adjectives and adverbs that characterized the qualities of items on the list. Finally, we asked them to analyze an organizational case study, and to begin their analysis by naming the situation (as a machine or as a plant or animal) with the phrase "This situation is like a _____," and in their analysis to use whatever images from the generated lists of adjectives and adverbs that they thought best represented the situation.

Our intention was to test whether a mechanistic versus organic metaphorical framing would result in different problem formulations. We were, of course, confident that it would since the power of a metaphor to frame a situation in a particular way (Schon 1979) and the use of organism versus mechanism had been so widely accepted in organization studies (Morgan 1986). Instead, we found that subjects continually surprised us with their ability to use either type of metaphorical framing to describe the case situation as being almost any kind of problem. Subjects could declare the organization to be too centralized or too decentralized, too rigid or too flexible, and could propose corrective restructurings based on those problem formulations, regardless of the metaphorical imagery they employed in their problem framing.

You would think at this point that we would give up on the idea of a frame as a cognitive construct with much power. Instead, we marveled at the human ability to play "language games" and were inspired by Wittgenstein's enduring insight into the multiplicity of the uses and meanings we humans make with the most familiar of words.

> 11. Think of the tools in a tool-box: there is a hammer, pliers, a saw,
> a screw-driver, a rule, a glue-pot, glue, nails and screws.—The
> function of words are as diverse as the functions of these objects.

Greenberg and I were not about to give up on the idea of reframing, though. If metaphors wouldn't reveal it, then perhaps being confronted with out of frame data would. We picked up and completed a project that I had begun with another colleague, Soong Park, a few years before (Boland, Greenberg, Park and Han 1990), in which subjects first analyzed a case study, then identified the alternative courses of action open to management and made a request for the data they would need in order to choose which alternative to recommend. Our manipulation was to either give the subjects the data they had requested or to give them data that had been requested by someone else, who was deciding between different alternative courses of action. Our expectation was that subjects who received the data they had requested would continue to operate within the problem frame they had initially developed, and would choose a course of action to recommend from among the original set they had identified. Subjects who received data they had not requested (data that was relevant to other ways of framing the problem) would tend to ask themselves the questions that the data helped to answer. We anticipated that they would be more likely to change problem frames and select a final course of action to recommend that was not one of the original ones they had identified. Once again, we were surprised by the results. Subjects who received data that was relevant to other ways of framing the problem clung tenaciously to the alternatives from their original problem definition, and selected one of those alternatives as their final recommendation. On the other hand, subjects who received the data they requested in order to choose among the alternatives they had identified frequently changed their problem frame and recommended a final course of action that was not in their original set of alternatives.

You would think that now we would start to give up on frame and try some other way of thinking about what is going on in peoples minds when they are thinking about organizational problems. But the idea of frame is just too powerful and obvious to give up. We kept interpreting these results as variations on the theme of framing, and argued that the results emphasized the creative, constructionist power people have to frame and reframe in the most surprising ways.

3. Map Making, Interpreting Numbers, and an Awareness of Narrative

Apart from these studies that were directly on the theme of framing, most of the work I have been involved in through the years has touched on related topics of sensemaking and interpretation (Boland 1984). I'll mention two of them here. One stream of work has to do with cognitive mapping. Ram Tenkasi, Dov Te'eni and I worked with new product development teams from DEC on the problem of reducing the "disconnects" that they were experiencing (Boland, Tenkasi and Te'eni 1994). Disconnects were when a new product came out and its consequences had not been adequately foreseen by various parts of the organization: the sales force had anticipated its uses or functionality to customers incorrectly, or the manufacturing facilities were inadequate, or the marketing plan didn't allow for key customer groups, or the financial projections were too high or too low. The project was to develop a cognitive mapping tool that would enable actors from different parts of the organization involved in the new product introduction to create, exchange and critique cognitive maps of their individual understandings of what the causes of success or failure of the new product were going to be. Our idea was that the problem of disconnects was at least in part a communication problem, and the different actors on these new product introduction teams each had different ways of framing their understanding of customer needs, alternatives in the market, moves by their competitors and so on. If they could each make the understandings from their different frames visible, we thought, they could develop better mutual understanding, be better able to take each others' frames of understanding into account and be able to reduce the number and scope of "disconnects" (Boland, Schwartz and Tenkasi 1992; Te'eni, Schwartz and Boland 1992).

We happened, unfortunately, to be working with DEC at the time of a great internal transition for them. The founder and CEO, Ken Olson, stepped aside, and a new management team and a more hierarchical style was put in place. The new product introduction process was changed drastically, and a comparison of before and after with the cause mapping communication process was not possible. Still, we did spend many months making and exchanging maps among the team members, and the results from that process and other mapping projects did teach us something about frames, maps and communication. We had begun the project with the idea that people did have frames and fairly well developed understandings of their areas of expertise. What we found was that individuals would start to make a cause map by putting a few factors and relationships into the map, and then stop. We had expected that people would be using the map as a way of recording their causal understandings, and worked on ways that the software tool would enable them to record them easily and flexibly, with rich linkages from elements in the map to underlying evidence and supporting data. Instead, we found that people didn't have much in the way of well developed understandings to put into the map in the first place. Making the map, in fact, almost seemed to be an opportunity to think about the causal relations as if for the first time. It was an opportunity for creative thinking that was invariably marked with expressions of discovery such as "Isn't this interesting," and "Look what's developing here." So instead of cause maps being a process of documentation of frames and exchange of understandings, they were a process of invention of frames in the first place and an almost naive exploration of these new found frames and their possible meanings.

This was probably the first dawning of the realization that perhaps the whole idea of people having "frames" was misguided. Frames, problem spaces, schemas all seemed so much harder to believe once you started working with people making maps of their area of expertise. They didn't seem to have a frame to simply report or document, but they could construct one and the experience of doing so was exciting and novel for them. So what was going on? What would it be like to do research of this sort that was openly "frameless"?

At about this same time, I began analyzing some data that had been collected by Ron Milne during his Ph.D. thesis, but not used by him (Milne 1981). He was interested in studying the effect of extrinsic versus intrinsic motivations on decision making in different budgeting contexts. Using a lab experiment, he had managers make choices about which of two managers to promote. In one condition, the managers had only actual accounting results for each manager's performance over the past three years. In the other condition, subjects had both actual and budget figures. Subjects had to chose which manager to promote, based on the accounting reports only, and were told that the candidates were equal on other, more personal dimensions. During his experiment, Milne had asked each subject for the reasons why they had chosen the manager they were recommending for promotion, but had not used that data in his analysis. I took that data and interpreted how those managers had interpreted the accounting reports in the two conditions (Boland 1993). I expected that the ways of making meaning with the reports would be different under the two conditions, and to an extent there were differences, but once again, my "real" finding was something unexpected. I argued that managers in both conditions made the accounting reports meaningful by using the numbers to bring the people behind them to life, endowing them with personalities, intentions, and histories, and animating their behavior in a hypothetical situation. Once

Table 1. Reasons for Promotion Choice

Reader A: no budget, choosing East

East appears to be innovative and resourceful. I would like the company to grow and I feel that East would be more likely to facilitate that end result. He more than achieved double the extra activities without doubling the cost. He takes risks within his department but they appear calculated to achieve outstanding results.

Reader C: budget, choosing East

The most important selection criteria was activities performed in addition to the standard 57. Typically, home office personnel departments get the tough problems bumped up from below. A good manager must be able to juggle a lot. Such a quantity of extra activities may indicate East is operating out of his department's scope—but it also indicates East is much more of a "go-getter" than West.

The expense data also indicated East was more aggressive—obtaining higher salaries and more equipment than West. Since such expenditures are normally approved by higher headquarters, East clearly has a pipeline to the right ears.

Least meaningful was the performance on the standard 57 activities. East is steadily doing better, but West has been consistently good. No other data was relevant. But personality and management style is very important for such a position.

Reader D: no budget, choosing West

I picked the "West" manager because he ran a tighter ship as far as budget was concerned and seemed to be more time organized. If he used the amount of money "East" did he would definitely have better results. Both had approximately the same number of employees but salaries were very different. "West" didn't spend an inordinate amount of time on activities other than the required since it is wasteful given the fact there is little direction as to what is important.

---

again, the way subjects were able to animate the people behind the numbers did not seem bound by the accounting results, or the presence or absence of a comparison to budget. They seemed to have a very potent interpretive capability, and could read the numbers as being good or bad as it pleased them. I positioned the analysis as a counterpoint to work in accounting which attributed a strong, common set of meanings to accounting reports (Macintosh and Scapens 1990).

Three examples of such differences in interpretation are shown in Table 1. Two things stand out relative to the purpose of this paper. First, the interpretation they make is dependent on the type of person they understand the manager to be, including the intentions, motivations, social skills, and personality they attribute to the manager. Second, they bring the person behind the numbers to life through the use of metaphor, but the metaphor is really a side issue compared to the way they narrativize the situation and the manager's behavior in it. The metaphor of a ship captain, an over achiever, or a juggler that they use is more of a "seed" for the telling of a story than a frame used for analysis. It was the motivated, intentional actor in a setting—facing a situation with a history and engaged in action with others who had their own interests—that was the medium of interpretation. It was the narrative that they constructed that propelled their reading and gave it meaning. And they were able to construct rich and diverse narratives from the scantiest of materials: columns of accounting numbers for department performance over the past three years. The bare numbers were enough for them to infer the kind of human agents and social setting that would produce such numbers, animate those agents in interaction with others, and "see what happened." It wasn't a schema or a frame which was at work, it was a story line that made the numbers sensible and the promotion choice inevitable.

The inspiration for taking narrative seriously, and for pushing the notion of frames aside, came from reading Jerome Bruner (1986, 1990). Early in his career he had been deeply involved in establishing the "information processing" school of cognitive psychology, and now he was arguing against the way that it had become a dominant, paradigmatic view of cognition. He was arguing that the information processing view was suppressing another, more frequently experienced mode of cognition that he was calling the narrative mode. In the information processing mode of cognition, we assess probabilities, we test hypotheses, we deduce consequences, and follow a pattern of "if x, then y" kind of thinking. In the narrative mode of cognition, we tell a story in which actors with causal powers enact a sequence of events, and follow a pattern of "first x happened and then y happened and then z happened" kind of thinking.

Through the information processing mode of cognition, we can show that we are right by demonstrating the logic of how our conclusions follow from our premises. In the narrative mode, we show we are right because the story is plausible and believable. The power of the story is in explaining why things are the way they are, and who we are as persons in a way that rings true to our experience and culture. In the information processing mode, the logic of our argument is supplied by the author and is apparent on the surface. In the narrative mode, the plot that ties the sequence of events together is supplied by the reader or hearer of the narrative as they subjunctivize its meaning for themselves.

The narrative mode of cognition and its portrayal of our situation as a plausible sequence of events is our principle way of making sense of our ongoing stream of experience. With it, we narrativize our life and its meaning to ourselves and others in real time as well as retrospectively and prospectively. Through narrative, we construct the identity of self and other as moral agents, assert what is true about our culture, repair apparent breeches in the canonicality of culture, and construct the conditions of our own action. If there is something like a frame or problem space within which the information processing mode of cognition operates, it is created through narrative. But somehow, narrative as a mode of cognition is consistently undervalued.

To take one instance of how, as Bruner argues, the information processing mode of cognition is given a predominant position in our world and of how the narrative mode is suppressed, consider Carol Gilligan's (1982) example of two children being tested with Kohlberg's hierarchy of moral values instrument. Jake and Amy are both 11 years old. As part of Kohlberg's standard test, they are told that Heinz has a wife who is dying and he cannot afford the drugs required to save her life. They are asked, "Should Heinz steal the drug?"

Jake resolved the problem by bringing it within a hierarchy of rights and responsibilities which enable him to use deductive logic. He said Heinz should steal the drug because a human life is more important than property. He in effect turned the problem into a simple kind of calculation, reflecting use of the paradigmatic or information processing mode of cognition, and he scored well on Kohlberg's instrument (between a three and four on Kohlberg's six stage scale).

Amy, on the other hand was unsure what Heinz should do. She tried to animate the situation and explore new possibilities:

> I think there might be another way besides stealing it, like if he should
> borrow the money or make a loan or something, but he really should
> not steal the drug. [p. 28]

When asked why he shouldn't steal the drug, she didn't use a calculation within a hierarchy of values, but continued narrativizing other possibilities for the unfolding situation:

> If he stole the drug he might save his wife then, but if he did he might
> have to go to jail, and then his wife might get sicker again, and he
> couldn" get more of the drug. [p. 28]

As Gilligan put it, Amy saw "in the dilemma not a math problem with humans but a narrative of relationships that extend over time" (p. 28). Amy did not employ the information processing mode of cognition to deal with Kohlberg's paradigmatically inspired theory, but instead employed a narrative mode to try and rewrite the story. The interviewer continued to ask her for an answer. Her stories were not codeable in Kohlberg's instrument. Eventually, Amy gave up and said simply that Heinz should not steal the drug "because it's not right." She then fit Kohlberg's scale and was scored between two and three, putting her below Jake in her moral reasoning.

Earlier, I had recounted Lou Pondy and I exploring the "symbolic" versus the "instrumental" mode of analysis. At that time, too, we saw the symbolic mode being devalued and pushed aside by the instrumental mode in organizational studies. Rather than retaliating by trying to deny the importance of the instrumental mode, we proposed that they formed a genuine union as an individual alternated back and forth between them, and that the symbolic mode played a role in framing and reframing the problem space used by the "instrumental" mode. Now, an awareness of narrative as a mode of cognition complicates that image. Bruner's information processing mode seems to match what we had in mind as instrumental forms of thinking, but I don't want to say that narrative is just a substitute for what we meant when we used the term symbolic, although clearly there is a connection of some kind. Eventually, I would like to explore how the narrative and the information processing modes of cognition form a genuine union, but in this paper, I would like to focus on how these different modes are being informed at a deep level by our basic concepts of space and time. I will propose that the information processing mode takes an inherently spatial approach to cognizing situations, and that the narrative mode takes an inherently temporal approach. As in the genuine union model, both are needed, but one dimension, the spatial dimension, has dominated and suppressed the temporal dimension. But first, let me give one example of a study of the narrative and information processing modes of cognition that I undertook with Ulrike Schultze (Boland and Schultze 1996), and then come back to this question.

During her study of the effects of information overload in computer mediated environments, Schultze collected examples of decision dialogues that took place in a Lotus Notes system at a large insurance company. The decision setting was a project called "Product Alignment" in which managers from around the United States were standardizing their rules for writing and pricing auto insurance policies. Below I will discuss some messages from that Product Alignment database:

Message 1:

Rule Number: 4
Key Person: S.A.
Rule Description: Driver assignment / FINAL

Policy premium is determined by assigning the highest rated driver to the highest rated vehicle, second highest driver to the second higher vehicle and so on. The highest rated driver refers to the operator whose age, sex, marital status and surcharges develop the highest premium. If there are more vehicles than drivers, rate each additional vehicle using a default class XX, where rates are based on the lowest rated liability driver class available on the policy with zero points.

Message 2:

Author: E.C.
Date: 09/08/93 09:44 AM

Comments at Sept 8 - 9 Evaluation

clarify rule to indicate that the lowest rated driver class on the policy (for liability) will be the one used for the default driver class, group agrees to use default driver class code XX.

These two messages come at the end of an extended process of discussions about rule 4, concerning the assignment of drivers to policies for pricing purposes. The first message gives a concise statement of the rule that the author sees as having emerged from their discussions. It is a strong information processing mode statement, and conforms to the image of cognition as movement through a problem space. Drivers, vehicles and their category descriptors are arrayed in the problem space and assignment is a logic for moving through the space: if X, then Y, else Z. In the second message, this mode is again evident as a sharpening of the categories that gives precision to the use of "default driver class." Both these messages have short, staccato phrasings, are assertive in tone, and use an abstract voice. They are intended to close the deliberations on this rule and depict it as a well formed problem space to be used by others. But in the next message of this data base we see the narrative mode of cognition used to reopen the search for meaning in their deliberations:

Message 3:

Author: R.C.
Date: 09/13/93 01:15 PM

Highest to highest, Not

Our rules say highest driver to highest vehicle, but what we calculate is highest total premium for the policy. These two can be different, here's an example:

2 cars, 2 drivers.
1 young driver with a bad record.
1 mid-aged driver (say a widowed woman) with a good record.
One car with liability only, symbol 80.
One car with full coverage, symbol 60.
Assume symbol 80 has slightly higher factors (for liability).

In this care "highest to highest" would probably put the kid on the symbol 80 vehicle. We all know the highest policy premium occurs when the kid is on the full coverage vehicle, which is the way we do it.

Over the years we've come to understand highest-to-highest as meaning the "highest total premium for a policy by aligning drivers and vehicles."

Message three disturbs the hoped for closure of rule 4 by using the narrative mode of cognition. The story, told in outline with a mocking introduction undermines the logic of the Final Rule by introducing characters that "we all know," putting them in action in a plausible scenario, and drawing a moral. We as readers subjunctivize the few narrative elements that are provided into a story. The author does not mention the sex of the "kid", but we know it's a male—and we can see him with a baseball cap turned backwards and the radio blaring as he drives too fast. We also know the female character. She is dressed modestly, talks in a soft voice and wears sensible shoes. "We all know" these characters and "we all know what we do" when we rate them. The narrative draws its power from the sense of canonicality and culture that it both draws upon and creates at the same time. The message is saying that our practices for dealing with these kinds of non-canonical situations and repairing the moral basis of our culture are who we are as a company. Narrating the rule in action in this way transforms it from a question of logic to a question of identity and assertion of self. The moral of the story is not stated explicitly, being left for the reader to complete as part of their interpretation. But importantly, this message has opened the issue for further discussion. What the information processing has closed down, the narrative mode has opened up.

Message 4:

Author: C.C.
Date: 09/16/93 12:08PM

Consistency

How we can have a "final" rule that does not conform to our practice? Are we affirming the rule and planning to change our alignment algorithm; or have we decided to file the rule and await a market conduct audit to demonstrate that we are not conforming to our filed rules?

Message four responds to the invitation of message three and extends the narrative mode. The author is moved by the story of message three and is exploring the apparent ethical dilemma presented by the rule as an expression of the identity of the firm. The author explores several paths the story could follow, and using an impartial narrative voice, opens the possibility for the reader to state a moral position: when rule and practice do not fit together in a coherent story, how do we proceed? Who are we and who will be if we live the story this way? What other story lines are open to us? These are some of the ways that the narrative mode of cognition can open new possibilities for action and create new opportunities for reflecting on the identity and culture of an organization.

4. The Spatialization of Thought and Experience

In this section, I want to explore some of the possible reasons and implications for the endurance of notions such as frame, schema or mental model. In my own experience of doing research, these notions have been prominent in both the construction and interpretation of studies. Yet, as I have tried to argue above, they don't seem to be supported by the results. This did not stop me from presuming the notions, using them, or acting in the name of them. And it doesn't seem to stop others either. The notion of frame or schema is something we just take for granted. In fact, it appears that Piaget is one of the few scholars to have done extended work on the creation and change of schemas over time, and even he didn't try to test the idea of schema itself. So where does the notion of frame or schema get its power and durability?

To begin with, the notion of frame or schema has roots that are very old and recurring in philosophy. Plato, as we know, argued that knowledge was based on ideal forms which are resident in the mind a priori. Particular ideas or experiences are understood as imperfect versions of the ideal forms already in the mind. Learning becomes a process of remembrance. In the dialogue with Meno, we see that geometry is known to the servant even without any formal education, if he is properly guided in the remembrance process. Aristotle, in contrast, argued that knowledge came from sensation and experience. Ideas were not ideal forms already in the mind but were the result of a process of stripping away inessential details from sensations, leaving abstract ideas. These two poles of the debate about human cognition seem to have been set very long ago, but either way, ideas such as frame and schema should have a ready place in our way of understanding cognition, because both shared the assumption of knowledge as an image in the mind.

In the enlightenment we again see these two poles. John Locke (1975) argued against the notion of a priori forms in the mind, emphasizing experience and sensation as the sources of all knowledge. The mind was a *tabula rasa* upon which the primary qualities of sense experience were recorded as simple ideas. The mind, in turn, built simple ideas into complex ones through the process of self reflection. As part of his argument against a priori forms, Locke pointed to the obvious lack of consensus on basic principles of political and economic organization. In contrast, Locke's contemporary, Gottfried Leibniz, argued for the necessity of innate ideas and a priori forms. But once again, both poles of the debate assumed that human understanding was based on an image.

Kant strengthened this image-based tradition when he drew in part on Leibniz and argued that we could know the world to the extent that it corresponded to the structure of mind, as we imposed a priori schemas to form images on the manifold of experience. Such schemas were forms of sensibility, lying ready in the mind.

> The schema of sensible concepts...is a product and, as it were, a monogram, of pure *a priori* imagination, through which, and in accordance with which, images themselves first become possible. [Kant 1965, A141-142 = B 181]

For Kant, the principle schemas were space and time, and of those, time was only understood by us when we could relate it to space. In his Inaugural Dissertation of 1770, he argued:

> The formal principles of the phenomenal universe which are absolutely primary, universal and at the same time the schemata and conditions of anything else in human sensuous cognition, are two, namely, time and space. [Kang 1985, p. 66]

> Space is employed as the type (analogy) even of the concept of time itself, representing it by a line and its limits—moments—by points. [Kang 1985, p. 67]

It's important to remember, though, that a schema for Kant was a constructive process.

> The schema is in itself always a product of imagination. Since, however, the synthesis of imagination aims at no special intuition, but

only at unity in the determination of sensibility, the schema has to be
distinguished from the image. [Kant 1965, A140 = B179]

Yet, it appears that the idea of schema or frame, to the extent that it is derived from Kant,
has come down to us in organization studies as a rather rigid, complete and well formed
image, rather than as the sensibility or the imaginatively constructed form that he had
intended.

Bartlett, who is most frequently seen as the direct source of most uses of schema and
frame in cognitive and organization studies today, appears to have suffered from the
same hardening of the construct at the hands of his contemporaries. Bartlett, in his
classic *Remembering: A Study in Experimental and Social Psychology*, complains about
the use made of his construct of schema:

> I strongly dislike the term "schema".... It suggests some persistent, but
> fragmentary, "form of arrangement," and it does not indicate what is
> very essential to the whole notion, that the organized mass results of
> past changes in position and posture are actively *doing* something all
> the time....It would probably be best to speak of "active, developing
> patterns"; but the word "pattern," too, being now very widely
> employed, has its own difficulties; and it, like "schema," suggests a
> greater articulation of detail than is normally found. I think probably
> the term "organized setting" approximates most closely and clearly
> the notion required. [Bartlett 1932, p. 200-201]

Later, reflecting on the difficulty of using the word schema to express his
experimental findings that remembering was an active construction, he concludes:

> It may be that what then emerges is an *attitude* toward the massed
> effects of a series of past reactions. Remembering is a constructive
> justification of this attitude; and because all that goes to the building
> of a "schema" has a chronological as well as a qualitative signifi-
> cance, what is remembered has its temporal mark; while the fact that
> it is operating with a diverse, organized mass, and not with single
> undiversified events or units, gives to remembering its inevitable
> associative character. [Bartlett 1932, p. 208]

Herbert Simon (1979, 1996) is perhaps the most well known exponent of the role
of a "problem space" in framing human problem solving. For Simon, the problem space
is always temporary, incomplete, and subject to revision, but none the less necessary.
In the problem space, there are nodes. Each node is an element of the problem or of a
possible solution. Problem solving is a movement through the problem space from the
current node to other, as yet unvisited nodes to test if they are taking the problem solver
closer to or further away from the solution or goal.

How is it that the schema or frame that has been handed down to us for so very long,
with each individual who is about to hand it off to us pleading for us to recognize the
inadequacy of the term, ends up being treated by us as if it were stable, well formed,
boundedly rational, coherent and suitable for navigation? Simon himself repeatedly

warns us that the problem space which results from any particular naming of the situation, or problem representation, is always inadequate—yet he proceeds to assume the existence of a well formed problem space and to conceive of problem solving as operating within it.

What is important to remember, I think, is that it is the temporal process of cognition in decision making or problem solving that is being modeled with the vocabulary of schemas and frames and problem spaces. Why is that? Why is the temporal experience being represented and modeled with spacial concepts? And does it make any difference? To me, the examples of the information processing mode that we saw before in the Lotus Notes messages is important to remember. It was used to shut the inquiry off, to close the problem space and end the discussion. The more temporally based narrative mode, on the other hand, was used to keep the question open and to keep the dialogue going. There must be a tradeoff between the opening and closing of inquiry, in which some balance between opening and closing is deemed most appropriate. But to the extent that spatial imagery and its associated information processing modes of cognition are treated as the only "real" cognitive account of problem solving, that balance is sure to be too heavily weighted toward the closing off of inquiry and dialogue. Which seems to be just the point when we complain about organizations not learning from errors, not being able to unlearn in the face of changing circumstances, and so on.

Henri Bergson (1911, 1920) had a very nice way of diagnosing this condition in which we seem to favor the spatial over the temporal in cognition. He makes a clear distinction between the way we can know the experience of time, and the way we can know objects in the extensible world. Objects can be measured and for practical (but ultimately arbitrary) reasons can be treated as separate or separable from other objects in their environment. Time, however, cannot be so measured because it is not made up of isolatable pieces and thus cannot be known by the same type of intelligence that knows objects. Time, he argued, is experienced as a duration in which there is no isolatable instant of "now." There is instead an ongoing experience of past, present and future interpenetrating each other as a continuous, seamless whole. Consider, as he does, the movement of your arm sweeping before you in space. As Kant had suggested, it can be known as a series of points in space only if we first imagine the trajectory it travels, and then, recounting the way the arm travelled through that trajectory, pretend that the experience of movement was like the line we have drawn in space. Like the line, our experience of movement, it is supposed, can be broken into separate pieces and measured in isolation as points in space (Bergson 1911, pp. 248-249, 290). But the experience of motion as a temporal duration has no such separate moments. It is an indivisible, undivided whole.

Bergson sees this same kind of confusion at work when we discuss the information processing view of cognition and describe the human problem solver as navigating a problem space following a branching model of problem solving. Temporally, the experience is like the movement of one's arm in that it is an undivided whole with each instant being an interpenetration of past, present and future. To talk of navigating problem spaces and following a path through nodes toward a goal is only possible in retrospect. In phrases presaging Weick (1979), Bergson critiques the spatial journey like description of cognition:

> Of course, when once the road has been traveled, we can glance over
> it, mark its direction, note this in psychological terms, and speak as if
> there had been pursuit of an end. Thus shall we speak to ourselves.
> But, the human mind could have nothing to say of the road which was
> going to be traveled, for the road has been created *pari passu* with the
> act of traveling over it, being nothing but the direction of this act
> itself. [Bergson 1920, p. 54]

5. Conclusions and Discussion

I have argued that the spacialization of knowledge and of our thinking about thinking is
a problem in that it tends to close off inquiry, to trap us into taking words like schema
and frame too seriously, and to lose sight of the temporal experience of meaning making.
It would be nice to be able to point to some simple remedy, but none seems available.
The dominance of space is ubiquitous in our language, in our day to day practices and
in our historically constructed ways of understanding cognition and organizations. It is
a tradition of thinking that has characterized both poles of the idealist-empiricist debate
throughout history. It is, according to Bergson, an inevitable feature of our attempt to
conceptualize temporal experience.

But that shouldn't make us give up trying. With that quixotic hope, one alternative
for a more temporal way of understanding meaning making in organizations would
involve a recognition of the narrative mode of cognition. But at present, the narrative
mode is systematically undervalued and even suppressed. Narrative is mere anecdote.
When it is studied, it is often spatialized as an object (Boje 1991). So at this point I can
only ask for giving more attention to the narrative mode without knowing exactly how
to do so. A key question is: can we develop more temporal methods for representing and
analyzing organizational phenomena? This suggests the need to design techniques of
representation and vocabularies for analysis that, like narrative, appreciate experience
as it unfolds, that are sensitive to rhythm, tempo and construction in the flow of
becoming.

There are others who sense a similar need to incorporate a temporal dynamic in their
work, and Fred Collopy, a colleague of mine at the Weatherhead School, is studying
some of them for insights into the design of temporal representations. These include the
early modern artists, such as Kandinsky and Klee, who sought for ways to give their
paintings the dynamic power of music. These also include avant guard artists such as
Morgan Russell and Stanton MacDonald-Wright, who worked on the design of music-
like instruments for playing visual images. In today's television and movies, we can see
numerous examples of a movement toward dynamic visual representations—from
computer graphics and computer animation to the daily weather reports. Without
knowing exactly how to do it, I suggest that we would benefit from developing tools for
representation in organization studies that are dynamic, multi-sensory, evocative, open
and tentative: tools that could create representations that are experienced with a sense
of organizational duration as the interpenetration of past, present and future.

In organizational studies, some examples of work that open up the narrative mode
of cognition and a more thoroughly temporal way of appreciating meaning making are
found in the study of discourse. Deidre Boden, in *The Business of Talk* (1994), for

instance, uses Garfinkel and ethnomethodology to understand how organizational actors in their micro-level "structures of practical action" talk selves and organizations into being "under pressing conditions of time and space" (p. 18):

> What is done now must make sense immediately, retrospectively, and to a consequential degree, in the future....as social actors discover *from within* the local logics of their actions, they reflexively...locate these activities within a stream of events that, taken together, constitute the rational production of the organization. [p. 198]

Boden, along with others in the discursive tradition, including Harre, Bilmes, Potter and Edwards, are giving us ways of approaching meaning and decision in organization as a lived experience, a *duree*, to be appreciated as a temporal whole.

Whether we think of the alternative to the information processing mode as a narrative mode of cognition, a discursive psychology, a dramaturgical analysis or an ethnomethodological analysis, the important thing is to appreciate that "Temporal change is the basic medium of all activity" (Waddington 1977). For understanding meaning making and decision, this temporal perspective highlights that context develops as action proceeds. Context is never simply available to resolve meanings in a problem space or to frame action with a branching model in an information processing sense.

We can imagine that actors have problem spaces and we can use information processing models to the extent that they prove handy while imagining things to be that way, but spatializing experience should not be allowed to suppress our search for other, more temporal ways of understanding organizational life. Actors as narrativizers of their own ongoing experience construct self and other as intentional, causal agents; read situations as they unfold; live in moments of tension that are only periodically and temporarily resolved; and struggle to recognize and repair breeches of canonicality. They are improvisational constructions of memory, meaning and action and it is that temporal medium of organizational life that we are at risk of losing as the spatialization of our knowledge about organization and cognition proceeds.

The objective is not to replace information processing modes of analysis, but, in the spirit of Rorty (1979, 1982), to make room for a temporal, narrative voice in hopes of keeping the conversation open and interesting. I hope together we can discuss possibilities for inventing tools of representation and vocabularies of analysis that might build a viable temporal presence in organizational studies.

References

Argyris, C. "Management Information Systems: The Challenge to Rationality and Emotionality," *Management Science*, 1971, pp. B275-292.

Bartlett, F. C. *Remembering: A Study in Experimental and Social Psychology.* Cambridge, England: Cambridge University Press, 1932.

Berger, P. L., and Luckman, T. *The Social Construction of Reality: A Treatise in the Sociology of Knowledge.* Garden City, NJ: Anchor, 1967.

Bergson, H. *Matter and Memory*, tr. N. M. Paul and W. S. Palmer. London: George Allen & Unwin Ltd., 1911.

Bergson, H. *Creative Evolution*, tr. by Arthur Mitchell. London: Macmillan & Co. Ltd., 1920.

Blumer, H. *Symbolic Interactionism: Perspective and Method.* Englewood Cliffs, NJ: Prentice-Hall Inc., 1969.

Boden, D. *The Business of Talk: Organizations in Action.* Cambridge: Polity Press, 1994.

Boje, D. M. "The Storytelling Organization: A Study of Story Performance in an Office-Supply Firm," *Administrative Science Quarterly,* 1991, pp. 106-126.

Boland, R. J. "The Process and Product of System Design," *Management Science* (24:9), 1978, pp. 887-898.

Boland, R. J. "Control Causality and Information System Requirements," *Accounting, Organizations and Society* (4:4), 1979, pp. 259-272.

Boland, R. J. "Sense-Making of Accounting Data as a Technique of Organizational Diagnosis," *Management Science* (30:7), 1984, pp. 868-882.

Boland, R. J. "Accounting and the Interpretive Act," *Accounting, Organizations and Society* (18:2/3), 1993, pp. 125-146.

Boland, R. J., and Greenberg, R. "Method and Metaphor in Organizational Analysis," *Accounting, Management and Information Technologies* (2:2), 1992, pp. 117-141.

Boland, R. J.; Greenberg, R. H.; Park, S. H.; and Han, I. "Mapping the Process of Problem Reformulation: Implications for Understanding Strategic Thought," in *Mapping Strategic Thought*, A. Huff (ed.). Chichester: John Wiley, 1990, pp. 195-226.

Boland, R. J., and Pondy, L. R. "Accounting in Organizations: Toward a Union of Rational and Natural Perspectives," *Accounting, Organizations and Society* (8:2/3), 1983, pp. 223-234.

Boland, R. J., and Pondy, L. R. "Micro Dynamics of a Budget Cutting Process: Modes, Models and Structure," *Accounting, Organizations and Society* (11:4/5), 1986, pp. 403-422.

Boland, R. J., and Schultze, U. "Narrating Accountability" in *Accountability: Power, Ethos and the Technologies of Managing*, R. Munro and J. Mouritsen (eds.). London: Thompson Press, 1996, pp. 62-81.

Boland, R. J.; Schwartz, D.; and Tenkasi, R. "Sharing Perspectives in Distributed Decision Making," *Association of Computing Machinery, Conference on Computer Supported Cooperative Work*, Toronto, 1992.

Boland, R. J.. and Tenkasi, R. V. "Perspective Making and Perspective Taking in Communities of Knowing," *Organization Science*, August 1995, pp. 350-372.

Boland, R. J.; Tenkasi, R. V.; and D. Te'eni "Designing Information Technology to Support Distributed Cognition," *Organization Science* (5:3), 1994, pp. 456-475.

Bruner, J. *Actual Minds, Possible Worlds.* Cambridge, MA: Harvard University Press, 1986.

Bruner, J. *Acts of Meaning.* Cambridge, MA: Harvard University Press, 1990.

Churchman, C. W., and Schainblatt, A. H. "The Researcher and the Manager: A Dialectic of Implementation," *Management Science* (11:4), 1965, pp. B69-B73.

Gilligan, C. *In a Different Voice: Psychological Theory and Women's Development.* Cambridge, MA: Harvard University Press, 1982.

Kang, Y. A. *Schema and Symbol: A Study in Kant's Doctrine of Schematism.* Amsterdam: Free University Press, 1985.

Kant, I. *Critique of Pure Reason*, tr. N. K. Smith. New York: St Martins Press, 1965 (1st ed., 1781, 2nd ed., 1787).

Kelley, H. H. "The Process of Causal Attribution," *American Psychologist*, February 1973, pp. 107-128.

Lakoff, G., and Johnson, M. *Metaphors We Live By.* Chicago: University of Chicago Press, 1980.

Leibniz, G. W. *New Essays on Human Understanding*, tr. & ed. Peter Remnant and Jonathon Bennett. Cambridge, England: Cambridge University Press, 1981 (1765).

Locke, J. *An Essay Concerning Human Understanding*, Peter H. Nidditch (ed.). Oxford: Oxford University Press, 1975 (Based on 4th ed., 1700, 1st ed., 1690).

Macintosh, N. B., and Scapens, R. W. "Structuration Theory in Management Accounting," *Accounting, Organizations and Society*, 1990, pp. 455-477.

Merleau-Ponty, M. *The Prose of the World.* Claude LeFort (ed.). Evanston, IL: Northwestern University Press, 1973.

Meyer, A. D. "Mingling Decision Making Metaphors," *Academy of Management Review,* 1984, pp. 6-17.

Milne, R. *Budget Slack,* unpublished Ph.D. dissertation, University of Illinois at Urbana-Champaign, 1981.

Morgan, G. *Images of Organizations.* Beverly Hills, CA: Sage Publications, 1986.

Rorty, R. *Philosophy and the Mirror of Nature.* Princeton, NJ: Princeton University Press, 1979.

Rorty, R. *Consequences of Pragmatism.* Minneapolis: University of Minnesota Press, 1982.

Schön, D. "Generative Metaphor: A Perspective on Problem Setting in Social Policy," in *Metaphor and Thought,* A. Ortony (ed.). Cambridge, England: Cambridge University Press, 1979, pp. 254-283.

Scott, W. R. "Developments in Organization Theory, 1960-1980," *American Behavioral Scientist,* 1981, pp. 407-422.

Simon, H. A. *Models of Thought.* New Haven, CT: Yale University Press, 1979.

Simon, H. A. *The Sciences of the Artificial,* 3rd ed. Cambridge, MA: MIT Press, 1996.

Sperber, D. *Rethinking Symbolism.* Cambridge, England: Cambridge University Press, 1974.

Te'eni, D.; Schwartz, D.; and Boland, R. J. "Cognitive Maps for Communication: Specifying Functionality and Usability," *Fourth Symposium on Human Factors in Information Systems,* Phoenix, AZ, 1992.

Waddington, C. H. *Tools for Thought.* London: Jonathan Cape, 1977.

Weick, K. *The Social Psychology of Organizing,* 2nd ed. Reading, MA: Addison-Wesley, 1979.

Wittgenstein, L. *Tracttatus Logico Philosophicus,* tr. D. F. Pears and B. F. McGuinnes. London: Routledge & Kegan Paul, 1961 (1921).

Wittgenstein, L. *Philosophical Investigations,* tr. G. E. M. Anscombe. Oxford: Basil Blackwell, 1974 (1956).

About the Author

Richard J. Boland, Jr. is Professor and Chair of the Department of Information Systems at the Weatherhead School of Management at Case Western Reserve University. Previously he was Professor of Accountancy at the University of Illinois at Urbana-Champaign. He has held a number of visiting positions, including the Eric Malmsten Professorship at the University of Gothenburg in Sweden in 1988-89, and the Arthur Andersen Distinguished Visiting Fellow at the Judge Institute of Management Studies at the University of Cambridge in 1995. His major area of research is the qualitative study of the design and use of information systems. Recent papers have concerned sense making in distributed cognition, hermeneutics applied to organizational texts, and narrative as a mode of cognition. Professor Boland is Editor-in-Chief of the research journal *Accounting, Management and Information Technologies,* and co-editor of the Wiley Series in Information Systems. He serves on the editorial board of six journals, including *Information.Systems Research* and *Accounting, Organizations and Society.*

5 DISTINCTIONS AMONG DIFFERENT TYPES OF GENERALIZING IN INFORMATION SYSTEMS RESEARCH

Richard Baskerville
Georgia State University
U.S.A.

Allen S. Lee
Virginia Commonwealth University
U.S.A.

Abstract

*It is incorrect and even harmful that many information systems researchers typically criticize their own intensive (qualitative, interpretive, critical, and case) research as lacking "generalizability." We untangle and distinguish the numerous concepts now confounded in the single term "generalizability," which are **generality, generalization, generalize, general,** and **generalizing.** These clarified terms allow us to identify four distinct forms of generalizing (everyday inductive generalizing, everyday deductive generalizing, academic inductive generalizing, and academic deductive generalizing), each of which we illustrate with an information systems-related example. The clarified terms provide the basis for an explanation of how information systems researchers who perform intensive research may properly lay claim to generality for their research.*

Keywords: Action research, case study, epistemology, experimental research, field study, intensive research, laboratory study, research methodology, statistical methods qualitative research.

For academic researchers in information systems (IS), the concept of "generalizability" has been developing and maturing with the growing acceptance of intensive research approaches (i.e., qualitative, interpretive, critical, and case research[1]). Once deemed a property of statistically based research alone, generalizability has been gaining recognition as an achievable ideal in intensive research as well (Lee 1989). Still, it remains common for authors of published intensive research articles to flagellate themselves in their own "discussion" sections for the lack of generalizability of their findings. There they typically blame this supposed failure on their having examined "only" a single case, or "only" three technologies, or "only" two organizations, or "only" one point in time, and so forth. We believe that such self-flagellation is not necessary (i.e., the completed intensive research can indeed claim "generalizability" if it is properly performed and presented).   Such unwarranted self-criticism can even be harmful (i.e., it can reflect and reproduce the hegemony of large-sample statistical research over all other forms of scholarly inquiry). The purpose of this study is to clarify the different processes of generalizing so that academic researchers in IS can better achieve and securely claim "generalizability" (or, as we will rename it, "generality") in their research.

The first section of the paper after this introduction will identify and distinguish different terms now confounded in the single term, "generalizability." The terms that we will unconfound and extract from "generalizability" are *generality*, *generalization*, *generalize*, *general*, *generalizing*, and even (after we define it) *generalizability* itself. In the section after that, we will define four types of generalizing and classify them according to the dimensions of "reasoning process" (inductive *vs.* deductive generalizing) and "context" (inquiry in everyday life *vs.* inquiry in academic research). The same section will illustrate the four types with empirical examples from the published literature. In the third section, we discuss an appropriate way in which information systems researchers may indeed lay claim to generality for their research. Then, in the last section, we will bring out the ramifications of the four distinct forms of generalizing for current and future research practices in IS. There we will indict and dismiss the dysfunctionality of self-flagellation for the often imagined sin of "lack of generalizability" in intensive research, as well as proclaim our emancipation to a research environment with a better (or "generalized," as it were) conception of "generalizability."

1. Unconfounding and Renaming "Generalizability"

In the way that academic researchers now use the term "generalizability," it regards the extent to which an academic researcher's theory does or does not apply to empirical referents (i.e., real-world situations) apart from the one that the researcher examined in his or her study. A case researcher, for example, might offer the self criticism (in the discussion section of her published case study) that the theory she developed "lacks

---

[1]We take this characterization of "intensive research" from M. Lynne Markus and Allen S. Lee's call for papers for a special issue of *MIS Quarterly* on intensive research. They, in turn, took the term "intensive research" from Karl Weick and used it to refer to the diversity of forms of empirical information systems research falling outside of the quantitative and positivist genre, including qualitative positivist (and non-positivist) research, interpretive research, critical social theory research, and case study research.

generalizability" because she based it on observations of only one information technology in only one organization. However, whether or not we would agree with her self criticism, we believe that she means *generality*, not *generalizability*.

In this study, we use the term *generalizability* more specifically. We distinguish it from *generality*, *generalization*, *generalize*, *general*, and *generalizing*. We define *generalizability* to refer to a theory's potential (as the suffix in *generaliz**ability*** signifies) to come to possess the quality of *generality*. *Generality* refers to the range of phenomena across which the theory has been demonstrated to hold. Here, *generality* is the outcome or product of the process of *generalizing* a theory that, initially, was able to be generalized (i.e., *generalizable*). To *generalize* is to engage in the process of *generalizing*. A theory of perfect *generality* would apply to the entire class of empirical referents that it purports to explain; a theory with less *generality* would apply only to a subset of this class.[2] In other words, we identify *generalizability* as a property of a theory at the beginning of an empirical investigation, and *generality* as a property of a theory at the end of the investigation (where the results of the investigation are favorable). The resulting *generalization*, in this scheme, would be the theory in the form in which it emerges from the empirical investigation. Finally, a theory possessing *generality* can also be described with the adjective *general*.

Methodologically speaking, *generalizability* is a potential that a theory has, just as *falsifiability* is a potential that a theory has. Just as (1) a researcher can take a newly formulated and untested theory that has the property of *falsifiability* (i.e., it is *falsifiable*, but has not yet been *falsified*) and then (2) the researcher, in examining the theory empirically, can *falsify* it and thereby demonstrate its *falsity*, we can say that (1) a researcher can take a newly formulated and untested theory that has the property of *generalizability* (i.e., it is *generalizable*, but has not yet been *generalized*) and then (2) the researcher, in examining it empirically, can demonstrate its *generality*.

Using the terms *generality*, *generalization*, *generalize*, *general*, *generalizing*, and *generalizable*, we now proceed to identify four types of generalizing.

2. Four Types of Generalizing

Our purpose in this study is to clarify the different processes of generalizing so that academic researchers in IS can better achieve and claim generality in their research. To this end, we offer a framework that draws attention to four types of generalizing.

First, we recognize that the act of generalizing is not something done only by academic researchers, but also something done by everyday people in everyday life. Consultants and managers, for instance, can (and, arguably, must) generalize from just one or two experiences. We therefore define one dimension of generalizing as referring to the context of inquiry. This context could be the inquiry of a person in everyday life (such as a consultant or a manager) or alternatively the inquiry of a person in academic

---

[2]We are aware of no published empirical study in the field of information systems that has ever offered *explicit* statements about what constitutes the entire class of empirical referents that its theory purports to explain. Even our own publications (i.e., the research that the authors of this paper have published) do not do this. It would be fair and justifiable to say that researchers generally imply what the class is (e.g., it could be "all managers" or "all corporations"). Still, we believe that, as a matter of good methodology, it would be a good idea for empirical studies to adopt this practice.

Table 1. Four Types of Generalizing

		Context	
		Inquiry in Everyday Life	**Inquiry in Academic Research**
Reasoning Process	**Induction**	*everyday inductive generalizing* "typification"	*academic inductive generalizing* "generalizability"
	Deduction	*everyday deductive generalizing* "learning"	*academic deductive generalizing* "falsification"

research (such as an IS researcher). In the context of inquiry in everyday life, the generality that a person associates with a belief is largely determined by the social traditions of the person's group. In contrast, in the context of inquiry in academic research, the influence of social traditions of the academic group in establishing generality are strongly mediated by rigorous notions of evidence, logic, and methodology in "scientific" thinking. Second, we recognize two different conceptions of the process of generalizing. These are inductive generalizing (a reasoning process that begins with observations and subsequently uses them as the basis on which to build a theory) and deductive generalizing (a reasoning process that begins with a theory and subsequently processes or tests it against observations). In Table 1, we use these two dimensions to identify four types of generalizing. We will now examine each of the four types in greater detail. Each type will be illustrated by means of an example. The examples were chosen not only because these illustrate the principles, but also because there are published details that illuminate the generalization process in each example.

2.1 Everyday Inductive Generalizing

The sociologist and phenomenologist Schutz offers an explanation of how everyday people in everyday life make generalizations.

> All projects of my forthcoming acts are based upon my knowledge at hand at the time of projecting. To this knowledge belongs my experience of previously performed acts which are typically similar to the projected one....The first action A' started within a set of circumstances C' and indeed brought about the state of affairs S'; the repeated action A'' starts in a set of circumstances C'' and is expected to bring about the state of affairs S''. [Schultz 1973, p. 20]

Regarding a person's general conception of a type of action, Schutz acknowledges that no two real-world instances or instantiations of it (e.g., A' and A'') are ever exactly the same, which leads to the following point regarding how a person in everyday life makes a generalization:

> Yet exactly those features which make them unique and irretrievable in the strict sense are—to my common-sense thinking—eliminated as being irrelevant for my purpose at hand. When making the idealization of "I-can-do-it-again" I am merely interested in the typicality of A, C, and S, all of them without primes. The construction consists, figuratively speaking, in the suppression of the primes as being irrelevant, and this, incidentally is characteristic of typifications of all kinds.

What Schutz calls a "typification" is further discussed by Berger and Luckmann in their classic book *The Social Construction of Reality* (1966). In one person's interactions with another person, the former could "apprehend the latter as 'a man,' 'a European,' 'a buyer,' 'a jovial type,' and so on" (Berger and Luckmann 1996, p. 31), where "these typifications ongoingly affect" how the former interacts with the latter. Typifications serve the purpose of allowing everyday people to negotiate their interactions with one another, as well as with the physical world around them. Returning to Schutz's example, we denote the process of generalizing in everyday life as follows:

$$A', A'', A''', \ldots ===> A$$

Another example, again referring to Schutz's typifications, would be:

$$\{A', C', S'\}, \{A'', C'', S''\}, \{A''', C''', S'''\}, \ldots ===> \{A, C, S\}$$

The reasoning process that produces such typifications is what we call "everyday inductive generalizing." We describe this reasoning process as "everyday" because it refers to inquiry in everyday life, rather than inquiry in academic research. We identify this reasoning process to be "inductive generalizing" because it begins with observations and subsequently uses them as the basis on which to generalize and then construct the typification.

Example of everyday inductive generalizing in IS

For an example, we will focus on one particularly interesting typification constructed by the professionals who work in the everyday world of IS development. It is what they call the "death march project."

They have come to see the "death march project" in the following way. It involves IS projects where signs of failure are apparent, but all participants in the project nonetheless proceed to play their parts, as if there were nothing wrong. "Death march projects" have come to be known, at least to those in the everyday world of IS development, by certain notable signs. One sign is that the risk of project failure appears greater than 50%. Other frequent signs of a "death march project" include irrational managerial compensation, the drowning of staff with a complete, sudden conversion to a new silver-bullet methodology, and the driving away of staff through cancellation of vacations and weekends (Yourdon 1997). Through everyday inductive generalizing across their observations of such IS development settings, software engineering practitioners have constructed the "death march project" typification. The sorts of projects now seen as "death march" projects are not new, having existed since the 1960s (Yourdon 1997); however, the typifying or generalizing of these observations into the shared, everyday concept of "death march" occurred only in the 1990s.

The process by which the "death march" typification results from the everyday inductive generalizing performed by a group of developers is illustrated in the February 1997 issue of *American Programmer*, which is dedicated to the topic of death march projects. In this issue, corporate IS practitioners and IS consultants speak normatively from their experiences with death march projects. Each death march project in their experience constitutes a discrete phenomenon that they have experienced, where P' could be the SMS/800 nine hour project (Oxley and Curtis 1997) and P'' could be the billing and accounts receivable system (Roberts 1997). Across these discrete experiences, they have generalized to P, the death march project:

$$P', P'', P''' ===> P$$

P, the "death march project" typification, is the general case. The process of generalizing commences by noting characteristics shared across the observed cases P', P'', P''', etc. Examples of these shared characteristics are certain irrational management actions noted above. The existence of the "death march project" was never "hypothesized" or "tested" in the scientific sense of these terms. Rather, this general case is a consequence of the generalizing by everyday IS developers across certain unpleasant projects. The general case is significant to these developers because of its instrumentality. The developers use the concept as a normative model to suggest possible actions when they encounter death march projects in the future. Although not tested for the status of scientific truth, the "death march" typification acquires the status of everyday truth "in so far as [it] helps us to get into satisfactory relation with other parts of our experience" (James 1975, p. 35).

2.2 Everyday Deductive Generalizing

Not all generalizations in everyday life are the "outputs" of a reasoning process for which particular instances are the "inputs." In other situations, an already existing generalization is itself the "input" to a reasoning process that applies it, where the "output" is the result that the generalization suited, or failed to suit, the application.

Argyris and Schön (1978) describe the processes involved in testing extant general theories in everyday life for suitability in new applications. They distinguish between a person's "espoused theory" and the same person's "theory-in-use." A person's espoused theory is the explanation that this person would voice to explain her behavior. A person's theory-in-use is the theory that actually governs this person's behavior. The theory that a person espouses is not always the same as the person's theory-in-use;

indeed, a person might not even be aware of her theory-in-use. When a person's actions, based on her theory-in-use, repeatedly elicit surprising reactions from other people or other parts of a person's environment—that is, when the theory-in-use that provided the basis for actions apparently does not work—an error or deficiency in the theory-in-use is sensed. Once this deficiency is sensed, the person can proceed to change her theory-in-use, her espoused theory, or both.

The everyday reasoning process in this behavior is not inductive, but deductive; it begins not with the observations of particulars, but with an already existing generalization (the person's theory-in-use). The person applies the generalization to a set of particulars she observes in the empirical setting in which she seeks to act (rather than derive the generalization from these particulars). Two results are possible. In one, the person notices surprises in the reactions to her actions, whereupon she learns that her theory-in-use needs to be changed (which would involve and require her becoming conscious of her theory-in-use for the first time). In the other result, there is nothing surprising for the person to notice, whereupon her theory-in-use survives and becomes, in a sense, further entrenched or stronger as a generalization. We refer to this reasoning process as "everyday deductive generalizing."

Everyday deductive generalizing is most apparent when conflict and "abnormal" discourse signal inconsistencies between the shared organizational espoused theories and the theories-in-use of individuals. In the process of this discourse, called "double-loop learning," the institutionalized generalizations are adjusted to match the individual experiences and theories-in-use. These corrected, espoused theories may then stand as general theories for further application.

Argyris and Schön were hardly the first scholars to recognize that deductive generalizing occurs in everyday life. For instance, this form of generalizing is the hallmark of the philosophy of pragmatism as found in the respective works of William James, John Dewey, and Charles Sanders Pierce.

Example of everyday deductive generalizing in IS

For an example, we will focus on the generalizing in which some people engaged in an IS action research project.

Action research is a research method that aims to solve immediate practical problems while expanding scientific knowledge. Based on collaboration between researchers and research subjects, action research is a cyclical process that builds learning about change in a given social system (Hult and Lennung 1980). Unlike *laboratory* experiments that isolate research subjects from the real world, action research involves *intervention* experiments in which the researchers, along with the research subjects, apply a stimulus or other change strategy to the real-world context in which the research subjects live or work. In action research, intervention experiments operate on problems or questions that the practitioners (to whom we also refer interchangeably as research subjects and everyday people) themselves perceive within the context of their own particular empirical setting. *Participatory* action research is distinguished by the additional characteristic of involvement of, first, practitioners not only as subjects but also as co-researchers and, second, researchers not only as scientists, but also as subjects. "It is based on the Lewinian proposition that causal inferences about the behavior of human beings are more likely to be valid and enactable when the human beings in question participate in building and testing them" (Argyris and Schön 1991, p. 86). Because of the deep collaboration between academics and everyday professionals, participatory action research involves both academic and everyday deductive generalizing. Consequently, examples of this form of research can illustrate the two categories of generalizing. However, in this portion of our study (section 2.2), we will focus on how the everyday professionals themselves engage in deductive generalizing. (We will defer our discussion on how the academic researchers engage in academic deductive generalizing until section 2.4.)

One example of an action research project in IS appears in Baskerville and Stage's study (1996) on risk management for prototyping. We may summarize their study as follows. It addresses one of the main problems in the overall practice of prototyping that, at least at the time of the study, had not yet been resolved. The practical problem was the difficulty simply in controlling the scope and unfolding development of prototyping projects. In general, this problem can severely limit the range of IS development projects in which prototyping can be used effectively. In Baskerville and Stage's study, the action research project developed and validated a new approach that uses an explicit risk mitigation model in the IS development process, one that focuses the collaborative action research team's attention on the consequences and priorities inherent in the prototyping situation. The study established that prototyping projects could be controlled if appropriate risk resolution strategies were put into effect prior to any breakdown in the prototyping process.

Baskerville and Stage's study illustrates how collaborative action team members engage in everyday deductive generalizing. In the example that follows, the generalizing pertains to the team members' original conception of a prototyping risk factor that they called "user alienation."

Specifically, while in the process of applying risk analysis as a tool to help administer their prototyping project, the collaborative action research team members stated certain expectations during an initial analysis of the risk factors that they perceived in the problem setting of their practical prototyping. Using the Argyris and Schön terminology that we introduced above, we can say that the collaborative action research team members came to "espouse" a theory that named 12 risk factors. At the same time, because they actually applied and followed their "espoused theory," we may state that, in this case, the "espoused theory" coincided with their "theory-in-use."

The collaborative action research team members stated one of the 12 risk factors as "Users will not understand what we are doing" (Baskerville and Stage 1996, p. 493) and they stated the consequence of this factor as "The users become alienated." They ranked this factor's risk level as moderate.

However, during the first prototyping cycle, the participants noticed that their understanding of risk factors led to a surprising result. The participants noted the surprise that the mentioned risk factor carried not just one, but actually two separate consequences. The first of the two separate consequences was enduring and serious: "The users do not know what product they will receive" (p. 495). The second was transient and trivial: "The users do not understand their role in the development process." The need to differentiate the original single consequence into two became obvious during the first learning cycle as the users gained experience with prototyping and adapted enthusiastically to their role in the development process (Baskerville and Stage 1991, p. 12).

To recapitulate, we note that the intervention experiment was not inductive, but deductive; the intervention experiment began not with the observations of particulars, but with a generalization about risk factors that the collaborative research action team members had previously formulated. As required in the invention phase of action research, the researchers were not observing as detached scientists, but instead were participating in the role of everyday people themselves who are acting and thinking in the real world. In their intervention, all of the team members applied their already existing generalization to a set of particulars that they observed in the IS development context in which they sought to act. In the results from their intervention, they noticed the major surprise in which there was not just one but two different consequences to one of their theorized risk factors, whereupon they learned that their understanding needed to be changed. We refer to this reasoning process as an instance of everyday deductive generalizing. Moreover, in this particular instance, the conclusion was not to claim greater generality for their initial understanding, but rather, to establish that it lacked generality. The everyday people *learned* that they needed to change their already existing generalization.[3]

2.3 Academic Inductive Generalizing

We observe that, among many (and sometimes, it seems like most) IS researchers, there is the (mis)conception that the greater the sample size, number of observations, or other quantity of empirical material in a research study, then the more "generalizable" (or, in the terminology of our study, the more general) the resulting theory is. We use the term "academic inductive generalizing" to refer to the process by which social scientific researchers would begin with observations (such as "n" sample points) and end up with a theory.

Lee (1999) reviews arguments, from the philosophy of science, that induction is fine for suggesting the formulation of a theory, but also that induction offers no help in testing or otherwise empirically validating a theory. The gist of his review is that (1) according to induction, the greater "n" is (where "n" is the number of observations across which a researcher would be generalizing a theory), the more general or "generalizable" the resulting theory is, but (2) the methodological principle that this procedure assumes—namely, that "inductive inference leads to valid theories"—is itself not empirically justifiably because (3) any attempt to provide an empirical justification of the statement, "inductive inference leads to empirically valid theories," would ultimately apply induction itself, thereby leading to an infinite regress in reasoning.[4] Lee

---

[3]Using the Schutz-based terminology of $\{A, C, S\}$, the conclusion was that $\{A, C, S\}$ lacked generality. Specifically, S did not fit and it needed to be replaced.

[4]Using the Schutz-based terminology, induction maintains that the greater the number of observations $(O', O'', O''',$ etc.), then the more general or "generalizable" the resulting theory, T. In other words, in the following,

$$O', O'', O''', O'''', \dots, O'''' ===> T$$

notes that even the discipline of statistical inference has distanced itself from the notion of induction, where statistical inference recognizes that a larger sample size, "n," does not increase the probability that a statistically inferred proposition (such as a confidence interval) is true, but instead serves to enhance the "level of confidence" or "statistical significance," which is an attribute describing the researcher's investigation, not the truth or falsity of the proposition.

As we will explain in section 2.4, deductive generalizing is the viable alternative to inductive generalizing in academic research.

Examples of academic inductive generalizing in IS

Although inductive generalizing is not a valid scientific procedure, many IS researchers hold themselves to it as the standard for establishing whether or not their research possesses what they call "generalizability" (or what we call generality). And almost always, they apply this standard (which they do not know to be incorrect) so as to conclude (incorrectly) that their research lacks "generalizability" (generality). We find instances of this when we look at IS researchers who conduct case studies. They regard the single case or small number of cases that they examined to dampen the "generalizability" (generality) of their results. Applying the same logic, they say that "generalizability" (generality) can result only by increasing "n."

> From the point of view of theory development, while case studies provide useful anecdotal information, the generalizability from one specific instance to another is often limited. [Albers, Agarwal and Tanniru 1994, p. 94]

> First and foremost, it should be reaffirmed that the single case research strategy employed here only allows generalizability to a research model, which in turn needs to be tested under a multiple case study design or by other field methods. [Brown 1997, pp. 90]

> From the evidence of the two cases, it was not possible to identify any generalisable [sic] strategies for overcoming constraints but the particular solutions developed appeared to reflect the developers'

---

the presumption is that as the number of observations (the instances on the left hand side of the arrow) increases, then the "generalizability" (generality) of the theory (on the right hand side of the arrow) also increases. The flaw is that the presumed statement, "as the number of observations increases, then the 'generalizability' (generality) of the theory also increases," is itself not empirically justifiable. For this insight, Lee credits Popper (1968) who, in turn, cites the philosopher Hume.

A not insignificant, additional point is that the presumed statement, "as the number of observations increases, then the 'generalizability' (generality) of the theory also increases," also begs the question of how it is possible to derive theoretical statements comprising T (which typically posit the existence of unobservable entities and relationships) from observation statements. If the many observations $O', O'', O''', O'''', \ldots, O''''^{..}$ can be generalized to anything, they would be generalized to O,

$$O', O'', O''', O'''', \ldots, O''''^{..} ===> O$$

rather than to T. This amounts to another flaw in the logic of induction regarding the empirical justification of a theory.

local conditions and their knowledge, intuition, and experience. This would suggest that rather than giving a set of generalized guidelines for improving user involvement (as is common in the literature), the emphasis might be better placed on supporting developers' ingenuity and improvisation and on developing their social skills to enable them to overcome the constraints on involvement. [Nandhakumar and Jones 1997, p. 84]

Because they are drawn from a study of two organizations, these results should not be generalized to other contexts. Each context is different, so we should expect different contextual elements to interact with technical initiatives to produce different consequences. The findings should not even be extended to other settings where GIS, or even Arc/Info, is implemented. What is true for GIS in the two local county governments studied may be untrue for GIS in other governmental units or in private enterprises. [Robey and Sahay 1996, p. 108]

In particular, in-depth analysis of extensive data from only one organization reduces generalizability, but increases correspondence to reality. [Hidding 1998, p. 311]

The study has a number [of] limitations that need to be considered in making any conclusions. First, the single case site limits the generalizability of results. The purpose of the study was not to provide generalizability of empirical results to other firms, rather the purpose was to "expand and generalize theories" (Yin 1984). [Jarvenpaa and Leidner 1997, p. 408]

Conducting additional case studies will provide instances of the various learning/outcome combinations, and we encourage such research. On the other hand, case studies alone will not result in validity or generalizability. Toward that end, a more fruitful approach might be to compare the development processes for similar systems in different organizations, or two or more systems being developed in a single organization. [Stein and Vandenbosch 1996]

The irony is that these IS researchers' method-in-use (with respect to generalizing) derives from a naïve and invalid statistical notion—the notion that a larger sample size leads to greater "generalizability" (generality)—to which even the field of statistics itself does not subscribe. However, once we understand the invalidity of this notion, we notice that the generality of the reported cases studies is inherently no different from the generality of statistically conducted, large-sample studies. As for IS researchers who flagellate themselves for a supposed lack of "generalizability" (generality) in their case studies, we believe that they are not only doing this unnecessarily, but also harming the overall reputation of the academic discipline of IS. This reputation is diminished in two ways: first, by reinforcing and perpetuating an incorrect notion of science (i.e., validity is achieved through induction) and, second, demeaning (and thereby discouraging the dissemination of) their own research findings whenever they draw public attention to a flaw that does not really exist.

2.4 Academic Deductive Generalizing

Lee (1999) also reviews arguments that explain how deductive inference operates in science. He notes that (1) the statements comprising a theory typically posit the existence of unobservable entities, such as molecules, atoms, electrons, and protons (hence making it appropriate for these statements to be called "theoretical"), (2) a researcher applies the theoretical statements to the facts describing a specific situation (such as the initial conditions in an experiment), allowing the deductions of predictions of what the researcher subsequently ought to observe or not to observe *if the theory is true*, upon the application of the experimental stimulus or treatment, and (3) the researcher, in comparing what the theory predicts in this specific situation and what she actually observes in this specific situation, ends up testing the theory deductively and indirectly.[5] If the prediction turns out to be false, then the theory from which it followed deductively would be false as well. On the other hand, if the prediction turns out to be true, then "the truth of the theory (from which the prediction originated) is not proven, but is only 'corroborated,' 'supported,' or 'confirmed' in the instance of this single test" because other new instances would open up the same theory to yet new opportunities for its falsification. Hence, in the deductive logic of empirical science, a theory can never be definitively shown to be true, but must remain forever falsifiable. Lee states, "the widespread characterization of theories, even in the social sciences, as falsifiable, testable, refutable or disconfirmable [is] an indication of the widespread extent to which the deductive testing of theories is practiced."

For academic deductive generalizing, unlike academic inductive generalizing, a larger number of observations consistent with the theory being studied does *not* increase the generality attributable to the theory. For instance, to increase the number of observations consistent with Newtonian physics is indeed possible, but hardly makes Newtonian physics more general (indeed, it is false and has been superseded by Einsteinian physics). Likewise, to increase the number of observations consistent with Ptolemaic astronomy is indeed possible, but hardly makes the theory of an earth-centered solar system more general! Therefore, in academic deductive generalizing, additional observations (such as a larger sample size) supportive of a theory *do not render the theory any more general or true.* This also means that the only observations that a deductive scientific researcher may regard as contributing useful information in gauging whether a theory is true or false are those that contradict the theory; such empirical

[5]Using the Schutz-based terminology, instead of having $\{A', C', S'\}$, where A' refers to an instantiated action, C' refers to an instantiated set of circumstances, and S' refers to an instantiated state of affairs that A' engenders in C', we have $\{T, C', S_p', S_o'\}$, where T refers to the scientist's theory, C' refers to the specific empirical conditions (the "initial conditions") instantiated in an experiment, S_p' refers to the specific state of affairs expected and predicted (hence the subscript $_p$ in S_p) to follow in the experiment if the theory is true, and S_o' refers to the specific state of affairs actually observed (hence the subscript $_o$ in S_o) to follow in the experiment. The theory T is falsified if S_p' and S_o' differ. The theory **T** is corroborated or confirmed if S_o' matches S_p', but remains falsifiable in future experiments involving the new circumstances, C'', and then C''', and then C'''', and so on.

evidence would, at best, throw a theory's validity and generality in doubt and, at worst, result in the researcher's conclusion that the theory is false and hence possesses no generality.

Example of academic deductive generalizing in IS

For an illustration of academic deductive generalizing, we return to the Baskerville and Stage action research study (1996) that we introduced in section 2.2. The example involves the detection of an important delusory element in the researchers' risk analysis. This element materialized when a risk factor, although highly expected, failed to develop as time progressed. The risk factor, "The technical environment is unreliable" (p. 493), was included because the prototyping project was forced to use a recent release of a database software package known sometimes to corrupt database files.

The software performed flawlessly in the early stages of the prototype construction. Based on this experience, the members of the collaborative action research team downgraded the probability of the risk factor in question. In this scenario, we can see that the researchers posited a hypothesis ("this software is unreliable") and conducted an experiment (prototype construction) that failed to support the hypothesis—hence (deceptively) encouraging them to attribute less generality to it. However, as the prototype development continued, its functionality and complexity rose. With this intensified usage, many previously unused features of the application generator were called into operation, with the result that the database software finally corrupted a table. This example thus demonstrates the action research equivalent to a Type II error, imputing support for a null hypothesis from the failure of experience to support the alternative hypothesis.

Since the time of their study's being published, the researchers have felt confidence in being able to apply their lesson beyond the particular action-research case is which it arose. They have formulated a general case in which prototyping management teams are inclined to reduce the probability rank of a known risk factor while the conditions that lead to such a risk became less favorable. Teams, in general, are theorized to rely on their own experience with risky aspects of projects, revealing a "guardian angel" mentality that inclines teams to ignore severely rising risk. The generality and confidence associated with this revised theory will remain ungauged until tested, like any other scientific theory.

3. Discussion

What is an appropriate way in which information systems researchers may lay claim to generality for their research?

First, and most important, information systems researchers should not give up claims to generality on the basis that their research involves a small "n" (for instance, a small number of organizations observed). The reason is that this would presume and reinforce the fallacy of the logic of induction—a logic rejected even by the natural-science model of research. A natural scientist who draws conclusions from an experiment does not rush to disclaim generality of her results on the basis that they are based on only a "single" experiment. In the same way, an information systems researcher who draws conclusions from a single field site need not rush to disclaim the generality of his results.

Second, the analogy to the experimental natural scientist also reveals how an information systems researcher may indeed lay claim to generality in her study. A natural scientist performing an experiment with favorable results (i.e., the theory's predictions are corroborated, not refuted) would need to state the details of the particular empirical circumstances for which the observed experimental results occurred. A claim of generality for the theory would mean that the theory can be expected also to hold in other instances *that share the same empirical circumstances.*[6] We note that an information systems researcher who conducts a case study (or other intensive research) would be as able to do this as a natural scientist who conducts an experiment. And just as the natural scientist conducting an experiment would not need to apologize for her theory's not necessarily applying to empirical circumstances different from those in her experiment, the information systems researcher who conducts a case study (or other intensive research) would not need to apologize for his theory's not necessarily applying to empirical circumstances different from those in his field site. Along the same lines, future efforts to enhance the generality of the information systems researcher's theory would not involve indiscriminately and randomly collecting more observations, but rather, would involve the targeting of additional specific empirical conditions to which the applicability of the theory is being questioned and, pending the observational results, would or would not be extended.

Returning to some of the published examples of empirical research in the information systems field that we presented in section 2.3, we suggest the following. We quoted Albers, Agarwal and Tanniru (1994, p. 94) as saying: "From the point of view of theory development, while case studies provide useful anecdotal information, the generalizability from one specific instance to another is often limited." Our suggestion is that Albers, Agarwal, and Tanniru can say instead: "From the point of view of theory development, a case study should not be characterized as providing anecdotal information, but instead should be valued, first, for concretely demonstrating a specific instantiation of the circumstances in which the developed theory is known to apply and, second, for allowing additional applications of the same theory to other situations also involving instantiations of the same circumstances. Specifically, these circumstances are...." Similarly, we quoted Robey and Sahay (1996, p. 108) as saying: "Because they are drawn from a study of two organizations, these results should not be generalized to other contexts. Each context is different, so we should expect different contextual elements to interact with technical initiatives to produce different consequences." Our suggestion is that Robey and Sahay can say instead: "Because they are drawn from a study of two organizations, these results can apply to other contexts sharing the same circumstances of these two organizations. Where the elements in other contexts are the same in these two organizations, we can expect interactions of these elements with technical initiatives to produce the same consequences. Specifically, these circum-

---

[6]Using the Schutz-based terminology, suppose we have $\{T, C', S_p', S_o'\}$, as we defined in the previous footnote. In the event that S_o' matches S_p', we can say that the theory T is sufficiently general for us to expect it to apply in all future occasions (whether future experimental settings or future organizational settings) that instantiate the set of circumstances C'.

stances are...." Finally, we quoted Hidding (1998, p. 311) as saying: "In particular, in-depth analysis of extensive data from only one organization reduces generalizability...." Our suggestion is that Hidding can say instead: "In particular, in-depth analysis from one organization enriches our knowledge of the fine details of the empirical circumstances in which the theory applies, hence increasing the reliability with which we can apply this theory in new settings in future occasions. Specifically, we can report the fine details of the empirical circumstances as comprising...."

Third, this notion of how to lay claim to generality also has an obvious ramification for practice. When an information systems researcher demonstrates that her theory applies in certain circumstances, a practitioner would know that these are the circumstances in which the theory could be used.[7] Where information systems researchers simply abandon claims of generality and hence forego specifying what these circumstances are, the transfer of academic findings to professional practice would be aborted.

4. Conclusion

The existing concept of "generalizability" is fastened on one peculiar form of generalizing. The concept is further confounded in the IS literature by its conflation with distinct concepts like generality and generalizing. Untangling this confusion reveals a variety of obscured, and sometimes legitimate forms, of generalizing. Many intensive research ventures could satisfy one of these forms, yet the authors self-flagellate over their inability to found their generalities on the one peculiar form that IS researchers have mindlessly idealized.

The self-flagellation is not merely annoying; it is harmful. When researchers unnecessarily divest their right to claim generality, their research audience is denied their analysis of the utility of their theories. When researchers generalize and claim generality, they encompass the larger scope of phenomena, beyond those directly captured by their research, to which their findings and understandings apply (Babbie 1990). By renouncing their right to generalize and claim generality, intensive researchers lose the latitude to explain the wide field of uses for their findings.

The IS research field is relatively vocational and operates in concert with technologies that are incredibly fast-moving. Our understanding of the social and organizational aspects of our field may be trailing far behind our grasp of the technical issues. Unnecessarily confining the application of new theories from intensive research is helping to cripple our ability to keep pace. Applying a variety of forms of generalizing in information systems promises to improve the development of more current and more useful social and organizational theory.

---

[7]Using the Schutz-based terminology, the same circumstances would be the C' in $\{T, C', S_P', S_O'\}$ for the information systems researcher engaging in academic generalizing, and the C' in $\{A', C', S\}$ for the practitioner engaging in everyday generalizing.

References

Albers, M.; Agarwal, R.; and Tanniru, M. "The Practice of Business Process Reengineering: Radical Planning and Incremental Implementation in an IS Organization," in *SIGCPR '94. Proceedings of the 1994 Computer Personnel Research Conference on Reinventing IS: Managing Information Technology in Changing Organizations.* Arlington, VA: ACM Press, 1994, pp. 87-96.

Argyris, C., and Schön, D. *Organizational Learning: A Theory of Action Perspective.* Reading, MA: Addison-Wesley, 1978.

Argyris, C., and Schön, D. "Participatory Action Research and Action Science Compared" in *Participatory Action Research,* W. F. Whyte (ed.). Newbury Park, NJ: Sage, 1991, pp. 85-96.

Babbie, E. *Survey Research Methods,* 2nd ed. Belmont, CA: Wadsworth, 1990.

Baskerville, R. L., and Stage, J. "Developing the Prototype Approach in Rapid Systems Modelling." Working Paper, Institute for Electonic Systems R91-35, The University of Aalborg, Aalbor, Denmark, September 1991.

Baskerville, R., and Stage, J. "Controlling Prototype Development Through Risk Analysis," *MIS Quarterly* (20:4), 1996, pp. 481-504.

Berger, P., and Luckmann, T. *The Social Construction of Reality: A Treatise in the Sociology of Knowledge.* New York: Anchor Press, 1966.

Brown, C. "Examining the Emergence of Hybrid IS Governance Solutions: Evidence from a Single Case Site," *Information Systems Research* (8:1), 1997, pp. 69-94.

Hidding, G. J. "Adoption of IS Development Methods Across Cultural Boundaries," in *Proceedings of the Nineteenth International Conference on Information Systems,* R. Hirschheim, M. Newman, and J. I. DeGross, (eds.), Helsinki, Finland, 1998, pp. 308-312.

Hult, M., and Lennung, S. "Towards a Definition of Action Research: A Note and Bibliography," *Journal of Management Studies* (17), 1980, pp. 241-250.

James, W. "What Pragmatism Means," in *Pragmatism,* W. James (ed.). Cambridge, MA: Harvard University Press, 1975, pp. 27-44.

Jarvenpaa, S. L., and Leidner, D. E. "An Information Company in Mexico: Extending the Resource-based View of the Firm," in *Proceedings of the Eighteenth International Conference on Information Systems,* K. Kumar and J. I. DeGross (eds.), Atlanta, GA, 1997, pp. 75-87.

Lee, A. S. "A Scientific Methodology for MIS Case Studies," *MIS Quarterly* (13:1), 1989, pp. 33-50.

Lee, A. S. "Researching MIS," in *Rethinking Management Information Systems: An Interdisciplinary Perspective,* W. L. Currie and R. D. Galliers (eds.). London: Oxford University Press, 1999.

Nandhakumar, J., and Jones, M. "Designing in the Dark: The Changing User-developer Relationship in Information Systems Development," in *Proceedings of the Eighteenth International Conference on Information Systems,* K. Kumar and J. I. DeGross (eds.), Atlanta, GA, 1997, pp. 75-87.

Oxley, D., and Curtis, B. "Reliable Work on a Death March Schedule," *American Programmer* (10:2), 1997, pp. 12-15.

Popper, K. R. *The Logic of Scientific Discovery.* New York: Harper Torchbooks, 1968.

Roberts, S. M. "Surviving and Succeeding in a Death March Project," *American Programmer* (10:2), 1997, pp. 12-15.

Robey, D., and Sahay, S. "Transforming Work Through Information Technology: A Comparative Case Study of Geographic Information Systems in County Government," *Information Systems Research* (7:1), 1996, pp. 93-110.

Schutz, A. "Common Sense and Scientific Interpretation of Human Action," in *Alfred Schutz, Collected Papers I: The Problem of Social Reality.* The Hague: Martinus Nijhoff, 1973.

Stein, E. W., and Vandenbosch, B. "Organizational Learning During Advanced System Development: Opportunities and Obstacles," *Journal of Management Information Systems* (13:2), 1996.

Yourdon, E. "Editorial." *American Programmer* (10:2), 1997, p. i-1.

About the Authors

Richard Baskerville is an Associate Professor in the Department of Computer Information Systems at Georgia State University. His research focuses on security and methods in information systems, their interaction with organizations and research methods. He is an associate editor of *The Information Systems Journal* and *MIS Quarterly*. Richard's practical and consulting experience includes advanced information system designs for the U.S. Defense and Energy Departments. He is chair of the IFIP Working Group 8.2, and a Chartered Engineer under the British Engineering Council. Richard holds M.Sc. and Ph.D. degrees from the London School of Economics. He can be reached at baskerville@acm.org.

Allen S. Lee is a Professor in the Department of Information Systems at Virginia Commonwealth University and the Eminent Scholar of the Information Systems Research Institute. He has been a proponent of qualitative, interpretive, and case research in the study of information technology in organizations. He is Editor-in-Chief of *MIS Quarterly* and co-editor of the book, *Information Systems and Qualitative Research* (London: Chapman & Hall, 1997). Allen can be reached by e-mail at AllenSLee@csi.com.

6 RESEARCH AND ETHICAL ISSUES ARISING FROM ETHNOGRAPHIC INTERVIEWS OF PATIENTS' REACTIONS TO AN INTELLIGENT INTERACTIVE TELEPHONE HEALTH BEHAVIOR ADVISOR

Bonnie Kaplan
Yale University School of Medicine
U.S.A.

Ramesh Farzanfar
Robert H. Freeman
Medical Information Systems Unit
Boston University
U.S.A.

Abstract

People develop strong reactions to machines. Among those reactions are personal relationships that people form with technological objects. Exploration of ways in which individuals relate to information technologies and how individuals see these technologies as being machine-like or being human-like is a fascinating and useful research area. This paper presents a study where these issues arose as a key, although unanticipated, theme in ethnographic interviews conducted among individuals using an intelligent interactive telephone system, Telephone-Linked Care (TLC), that provided counseling about health behaviors. Many interviewees expressed ambiguity over whether they were talking to a machine or to a person during TLC conversations. Interview findings suggest that people formed personal relationships with the TLC system, or at least with the voice on the telephone. These relationships ranged from feeling guilty about their diet or exercise behavior to feeling love for the voice.

Relationship formation with information technology is of special concern in areas such as medicine and health care in light of the debate over whether computers dehumanize patients. However, this concern pertains to other applications of information technology as well. The findings in this study raise ethical issues and the need for further research.

The value of ethnographic interviewing also is illustrated by this study. Ethnographic interviewing allowed new ideas–in this case, people's relationship formation with information technology–to emerge and, thereby, provided the possibility for future studies and the potential development of theory.

Keywords: Social and organizational issues, ethical issues, qualitative research, social informatics, medical and health care informatics, information systems evaluation, ethnographic interviewing, personal relationships with technology.

1. Introduction

It has long been recognized that people form relationships with technological objects. They can become attached to their automobiles (Rae 1965), for example, or they may see dynamos as symbols of an age (Adams 1918). A propensity for projecting human qualities onto things is apparent when people anthropomorphize computer technology. As psychologist Robert F. Bayles observed, "The computer is the ultimate Rorschach test" (quoted in Nelson 1974, p. 9). This observation was fleshed out in the work of Sherry Turkle (1982, 1984). She described how people project onto computers their own aspirations, ideologies, hopes, and politics. She argued that people of all ages, from school children to researchers in artificial intelligence at the Massachusetts Institute of Technology, use computers and related technology, such as video games, in ways that explore the boundaries between human and machine. In so doing, the people Turkle studied considered what it means to be human by thinking about what differentiates a person from a machine such as a computer. Computers, she explains, are evocative objects, well-suited to think with about such questions. They are just human-like enough to raise difficult questions, such as whether a machine can cheat. They are protean enough to serve as projective devices, much like the Rorschach ink blots, or, as Turkle described, other technological objects such as gears or mirrors. Mechanical dolls and similar automatons are both fascinating and disturbing for much the same reasons as are computers: they share similarities and differences with humans.

The issue of what human qualities are shared by computers, and what implications this poses, is not new. Alan Turing raised it in a famous paper as early as 1950 in which he proposed what came to be called the Turing Test as a way of answering the question, "Can Machines Think?" (Turing 1950—this paper has been published and reprinted under various titles). Since then, there has been a distinguished philosophical literature concerning this question, while new fields of psychology developed in which human mental functioning is explained in information processing terms. Zuboff (1988) raised related issues when considering how to design and use information technologies that replace human mental labor when she explored implications of automating vs. "informating" work. There has been extensive concern in the sociological literature, and some in the Information Systems literature, about how information technology can support work or dehumanize individuals.

Such questions become particularly important when information technologies are used in personal areas such as health care. As patient-centered health care grows in popularity, it is becoming more common place for information technology to be used to collect data directly from patients and to provide educational information to them. Such systems have been evaluated in terms of quality of data gathered from patients (Hunt et al. 1997), how satisfied patients are with such methods (Kim et al. 1997), and health outcomes for patients who use such systems (Balas, Boren, and Griffing 1998). Little has been done, however, to understand patients' reactions to such systems (Forsythe 1996). Concern over patient reactions is especially important in light of the debate over whether computers dehumanize patients (Shortliffe 1994).

Ways in which individuals relate to information technologies and how individuals see these technologies as being machine-like or being human-like is a fascinating area that needs more exploration. This paper presents a study where this issue arose as a key theme in ethnographic interviews conducted among individuals using an intelligent interactive telephone system that provided counseling about health behaviors. Although it was not an initial goal of the study, interviewees unexpectedly expressed ideas and feelings that suggest that they formed strong personal relationships with the system. They identified with the system in various ways and projected deep personal concerns onto it, while also expressing ambiguity and ambivalence over whether they were dealing with a person or a machine.

The value of ethnographic interviewing also is illustrated by this study. That so interesting and powerful a theme arose when it was unanticipated indicates that the study could not have been designed to explicitly address this issue. Ethnographic interviewing allowed new ideas to emerge during the course of the study and, thereby, provided the possibility for future studies and the potential development of theory (Glaser and Strauss 1967; Kaplan and Maxwell 1994).

2. Telephone-Linked Care

The Telephone-Linked Care (TLC) system is an intelligent interactive telephone system that uses interactive voice response technology to conduct a series of brief (approximately five minute) conversations with patients in their homes over their telephones. TLC uses a pre-recorded digitized human voice to speak to users. It can ask questions; respond to users' answers to those questions; and provide information, advice, and counseling. TLC has been used for a variety of health purposes, including disease monitoring and health behavior counseling (Friedman 1998; Friedman et al. 1997; Friedman et al. 1998). TLC-HealthCall is a TLC application that advises individuals on improving their health either through changes in diet or exercise, depending upon which version of the system an individual used. TLC spoke to users according to a script that was recorded by a human male voice professional. Participants called the system once weekly for a maximum of 26 weeks. By using the telephone keypad for input, they provided information in answer to questions TLC asked them. TLC then responded to this information with comparisons to their previous behavior and suggestions for future behavior. For example, TLC might have asked someone how much exercise he had had that week, or how many servings of vegetables she had eaten. In response to the person's response, TLC might praise the individual for increasing the amount of exercise done, urge the person to eat more vegetables, and provide suggestions for how to incorporate vegetables into the diet.

3. Methods

3.1 Interviewee Selection

An evaluation of TLC-HealthCall is being conducted to investigate a variety of health related questions. One aspect of this research includes a multi-method assessment of individuals' reactions to using TLC (Kaplan 1997; Kaplan and Duchon 1988). Among the evaluation research questions were what patients thought about using the system, whether they were satisfied with it, how it actually helped them, and what meaning individuals made out of their experience with TLC. Ethnographic or qualitative methods are especially useful for this kind of evaluation (Kaplan and Maxwell 1994).

The total number of individuals who completed the TLC-HealthCall study was 247. Of them, 215 were women and 83 were men. From these 247 people, 15 were interviewed in depth during a one-week period after they had completed 26 weeks of potential TLC use. Interviewees were selected so as to include:

  (1) individuals representative of the entire population in the study,
  (2) individuals who represented outliers of specific interest to the study, and
  (3) individuals who had the potential to be helpful informants.

All interviewees had completed an exit telephone survey with a member of the research staff. The survey protocol was a lengthy structured questionnaire that included, among other items, questions about respondents' attitudes toward TLC and its usefulness. Six of the individuals involved in the in-depth interviewing were chosen on the basis of their responses to these attitudinal survey questions. In particular, respondents were selected based on their answers to the three open-ended survey questions inquiring about the benefits gained from using TLC, suggestions they had for improving it, and a request for any additional comments. Their responses were reviewed so as to insure including articulate people who expressed positive attitudes, or negative attitudes, or a mixture of both. The remaining nine interviewees were selected at random in order to make the interview sample more representative. This process of mixed selection and randomized interviewees was intended to insure a broad representation in the interview sample. Because no attempt was made to ensure gender balance in the interview sample, it turned out that all but two of the interviewees were women. The study design allowed for additional interviews if saturation were not reached in the first set of interviews.

3.2 Interview Data Collection

Individuals who were recruited for in-depth interviewing were offered a $5 phone card as a gift in appreciation of their participation in this phase of the TLC study. Interviews were conducted in person by the first author in peoples' homes, work places, or our offices, depending upon the preference of the interviewee.   The second author participated in 12 of the interviews. Interviews lasted between 45 and 60 minutes. All interviews were tape recorded, with the interviewees' permission. In addition, both researchers hand-wrote detailed interview notes during the interviews. As explained to interviewees, tapes were intended as back-up to written notes, if needed. Tapes were not transcribed.

Ethnographic interviews were conducted in order to elicit how interviewees conceptualized, reacted to, and attributed meaning to their TLC experience. Consequently, interviews were open-ended so as to probe what interviewees said without constraining them in topic or response. Interviewees were told the purpose of the study, assured of confidentiality, and then asked to describe their use of TLC and what a TLC session was like. If the following topics were not brought up by the interviewee, interviewers asked for interviewees' assessment of benefits they had derived from using TLC, for their reaction to the voice, whether they had received anything in the mail (to explore the usefulness of periodic reports they were mailed), whether they had discussed anything about the study with their doctors, what suggestions they had for TLC *per se*, what they thought of other possible uses of TLC in health care, what they would tell a family member about TLC, what they thought of the training, and why they called the system as frequently or infrequently as they did. Some participants also were asked whether anyone else in their families had been involved in the TLC conversations. In the course of discussing these topics, interviewees spoke of changes that had occurred in their diets and exercise patterns and of their feelings about using TLC. Most interviewees were very forthcoming and comfortable talking about these topics.

3.3 Interview Data Analysis

The authors discussed interview findings during and after the week of interviewing. The first author, who was present at all of the interviews, analyzed her hand-written notes according to four standard methods for analyzing qualitative data: coding, analytic memos, displays, and contextual and narrative analysis (Kaplan and Maxwell 1994). She then compiled a detailed report that provided an analysis and discussion of results. This report was reviewed by the authors and additional members of the TLC project team as a step in insuring validity of results. What follows includes quoted remarks by interviewees and, even when not quoted, an attempt to capture as much of the interviewees' wording as possible (where verbatim quotations were not recorded, much of the wording was), based on the first author's interview notes.

4. Findings

Findings are reported for 14 of the 15 individuals who were interviewed. One of the two men who was interviewed was eliminated from interview data analysis because he claimed to have called into the system regularly and explained circumstances of all these calls. It subsequently was found from system logs that, in fact, he had called only once. Because we were concerned that his responses were fabricated, we discarded this interview from analysis.

To aid in interpreting the interviews, we identified several key themes by using a grounded theory approach (Glaser and Strauss,1967; Strauss and Corbin 1990). This paper concentrates on one striking theme from the interviews: how participants related to and formed relationships with TLC. This theme was unanticipated. Interviewees were not asked explicitly about relationships they formed with TLC and they had not made comments about it during the survey.

People expressed both positive and negative attitudes toward TLC. Positive reactions included that TLC was "fun," made them more aware of what they ate or how much exercise they did, gave them helpful information, served as a friend or mentor, and helped them change toward a healthier life style. Negative reactions were that TLC talked down to them, treated them like a child, made them feel guilty, had an unpleasant or "disembodied" voice, was inflexible and did not allow them to get or input information they thought important, was boring and repetitive, and did not help them. These responses entailed expression of feelings of guilt, love, and ambiguity or ambivalence, all aspects of forming a personal relationship with the system.

4.1 Guilt

The first indication of personal relationship formation emerged during the interviewing when interviewees expressed feelings of guilt evoked by TLC. This finding is related to negative ways in which they described the tone of TLC conversations. Some study participants described the voice they heard on the phone negatively, as a "disembodied" voice from on high or "a man with a tracheotomy." Some thought the voice was synthesized and were surprised to find that a person had made the recordings. As these remarks suggest, they not only thought the voice quality unpleasant, but also felt the tone was problematic. The problematic aspects arose when participants talked about how condescending and condemning TLC was. They felt they were being talked down to, much as their mothers had chided them when they were children. One woman said that she started lying to TLC in order to avoid the lectures she got from it. There was no allowance, they reported, for what they considered reasonable deviations from the behavior TLC was recommending. Nine people wanted to explain to TLC why they could not follow its advice, or they felt that TLC was not flexible enough to take account of their situations. Two individuals wanted to be able to leave messages and three wanted to record comments. People felt constrained by not being able to explain that, for example, they could not walk because there was a blizzard, or could not eat fish because they could not afford it. It seems that these participants felt a need to communicate more than just yes/no or numeric responses to TLC questions. They said they felt guilty about their behavior, or that they were being faulted for something they had reason to have done (or not done).

In addition to the desire to make explanations was the feeling among some interviewees that they were unable to provide information to TLC that would make their inputs more accurate. For example, one woman told of having eaten three donuts over the weekend. Because she called on Monday, and TLC asked only about the previous three days, she could input that she had eaten only three donuts. However, she had really eaten five donuts over the past week, and she wanted to have been able to input that. Mixed in with her apparent concern for accuracy was that she spoke in ways suggesting that she wanted to confess about the additional donuts, raising the idea of her guilt at having eaten them.

These findings of problematic tone and the need to explain and confess go beyond people's sense that the system was rigid and did not allow them to provide explanations or information they wished they could tell it. What is striking is that they felt a need to provide these explanations and information at all, that they cared about what they told

TLC. This went beyond their sense of responsibility to provide accurate information to what interviewees thought of as a scientific study. The reaction seemed to involve feelings of guilt.

Five people specifically brought up the issue of guilt and said that TLC made them feel guilty. Others said that when they ate the wrong things, they wanted to tell TLC. There was a sense in their comments that they had transgressed and needed to confess, like the woman who had wanted to tell TLC about the extra donuts. As another woman remarked, "I'm Catholic. I know about guilt." Her description of TLC as a "disembodied voice" that came from on high may have added to a sense of sins being known and needing expiation. Others, while not expressing feelings of guilt, brought up issues that may be related: TLC's acting like a conscience, their need to make explanations to TLC, their concerns with accuracy, and their embarrassment at being, as one person described herself, "a loser." All these seem related to some individuals' explicit concern with impression management and ways to look better to TLC. For example, several discussed how they wanted to have their conversations on days when they could report food consumption that would be more in line with TLC recommendations instead of on days (like over the weekend) when they might have eaten too much unhealthy food.

4.2 Love

At another end of the reaction spectrum from expressions of guilt were expressions of love. Three women talked in glowing terms about TLC, how it had served as an "unseen friend, a conscience" and a "helper...and mentor." All three said that they continued to call after the end of the study so they could continue the relationship. They spoke of a sense of loss at the abrupt ending of the TLC relationship. One explained it, "I missed that voice, having him say, 'Good morning.' I liked it." Another said she missed him because she got "trained, like Pavlov's dogs." The third felt that the machine replaced her family. It was someone to tell her what to do, like a part of her family.

Two of these people wanted to meet the person whose voice it was. One explained this by saying she wanted to meet the "doctor" on the phone to find out the philosophy of the program and because she "needed human contact." The other spoke in the same way as she might have if she wanted to meet a man who had sounded friendly, attractive, and appealing when he called, for example, by mistake or to conduct a phone survey or had in some other way had made contact without her knowing who he was. She sounded enamored as she said, "Oh, I love the voice. I'd love to meet the man with the voice." These women talked in terms and spoke in tones that gave the sense they had fallen in love with the man whose voice they heard.

4.3 Ambiguity and Ambivalence

The forming of personal relationships arose in another way as well. Many participants, regardless of how enthusiastic they were about TLC, referred to the voice as "he" and did not speak of "TLC' or "the system" but, rather, described what "he" said. However, there was ambivalence about whether TLC was a person or a machine. Interviewees differed in their sense of whether they were talking to a machine or to a person when using TLC. There was ambiguity about this, even for the same interviewee, with some calling TLC "him" while also referring to "it" and saying that TLC was a machine.

Six people talked explicitly of how TLC was a machine and sounded like one. They all seemed negative about the experience and indicated that they would have preferred dealing with a person. Four people made a connection between how people reacted to TLC and how they personally react to machines they consider similar, such as answering machines or automatic teller machines, which they did not like using.

On the other hand, five people explicitly spoke of TLC as though they were speaking of a person. As one put it, "This is my helper. This is my mentor." Four of them described both a personal relationship with TLC and also ways it was like a machine. Two other people compared TLC to a person and, specifically, to family. One person spoke of TLC as a substitute for someone in her family, someone to tell her things. She laughed as she said, "I get attached to a machine. It's like a person." As this remark suggests, there was some ambiguity in whether or not interviewees considered "him" a person or a machine. Interviewees indicated that there were advantages in having it both ways.

One advantage of this ambivalence arose when two women discussed such personal issues as loneliness, self-esteem, and depression. They related these issues to their interaction with TLC in ways that sounded as though TLC could function as a psychiatrist or psychological counselor for them. The anonymity of interacting with a machine apparently helped them do this. One of these women was very positive about her experience with TLC. She was the woman who thought of TLC as a family member. The other, who was negative toward TLC, spoke of how it could provide an outlet for something akin to counseling or friendship. "A lot of people don't have anyone to talk to. A computer is anonymous," she said, and much later in the interview, she remarked, "It shouldn't be a psychic hot-line."

5. Discussion: Personal Relationship

Despite the ambiguity and ambivalence described above, on the whole, people spoke of their interaction with TLC in the same kind of terms as they would have used if they were describing interaction with a person. Many indicated that they would have preferred human contact to that of a machine. Even so, even those who felt TLC was "disembodied," even those who said they wished they had been able to talk to a person instead of to the recorded voice, referred to "him." It seems as though participants were forming personal relationships with TLC, or, at least, with the person whose voice they heard.

This interpretation provides a connection between those who felt guilt and those who felt love. Although at different ends of the reaction spectrum, guilt and love both involve personal relationships. In both cases, as well as in expressions of ambiguity and ambivalence, these feelings were evoked by the relationship people formed with TLC. As with the women who thought of using TLC as a possible psychological encounter, interviewees who expressed guilt or love seemed to identify with TLC, and to project onto it human emotion and a personal relationship.

6. Conclusions

That people formed personal relationships with a system such as TLC raises important issues. Profoundly felt ethical concerns accompanied development of computer systems (e.g. Weiner 1964). When using this technology for artificial intelligence, issues can become so threatening as to be actively suppressed (see, e.g., Dreyfus and Dreyfus 1986). Weizenbaum (1976) was profoundly disturbed that people felt so personally involved in his ELIZA program, which simulated a Rogerian therapist, that they requested privacy while using the computer. An incident of this kind with his secretary provoked him to consider the ethical implications of computer use and resulted in his powerful *crie de coeur* that we, as information systems developers, implementers, users, and advocates, should consider the morality of what we do.

Turkle (1984) described numerous ways in which people project personalities and values onto computers and how people develop personal relationships with computer systems. As computer systems became more fluid, she presented new studies of how people take on different identities through computer use (Turkle 1995). Her work illustrates powerful ways in which information technology raises issues of identity and psychological projection.

The TLC study suggests that people had reactions to a computer-based simulated telephone conversation that are similar to those Turkle described. Individuals appeared to form personal relationships and identify with TLC, and to describe TLC in human-like terms, where their interface with this computer system was through a pre-recorded telephone voice and telephone key-pad. These findings raise both research and ethical issues. Among questions to be considered for future study are:

(1) What accounts for the kind of relationship individuals formed with TLC?
(2) How was this relationship affected by the nature of the interface, e.g., a computer-mediated conversation vs. direct use of a computer keyboard and display terminal?
(3) What ethical considerations are involved in creating or promoting systems with which people form deep personal ties?
(4) What implications does this personal relationship formation have for systems design implementation, and research?

References

Adams H. *The Education of Henry Adams.* Boston: Houghton Mifflin Co., 1918, 1961.

Balas, E. A.; Boren, S. A., and Griffing, G. "Computerized Management of Diabetes: A Synthesis of Controlled Trials," in *Proceedings of the AMIA Fall Symposium*, C. G. Chute (ed.). Philadelphia: Hanley & Belfus, Inc., 1998, pp. 295-299.

Dreyfus, H. L., and Dreyfus, S. E. *Mind Over Machine: The Power of Human Intuition and Expertise in the Era of the Computer.* New York: The Free Press, 1986.

Forsythe, D. "New Bottles, Old Wine: Hidden Cultural Assumptions in a Computerized Explanation System for Migraine Sufferers," *Medical Anthropology Quarterly* (10), 1996, pp. 551-574.

Friedman, R. H. "Automated Telephone Conversations to Assess Health Behavior and Deliver Behavioral Interventions," *Journal of Medical Systems* (22), 1998, pp. 95-101.

Friedman, R. H.; Stollerman, J. E.; Mahoney, D. M.; and Rozenblyum, L. "The Virtual Visit: Using Telecommunications Technology to Take Care of Patients," *Journal of the American Medical Informatics Association* (4), 1997, pp. 413-425.

Friedman, R. H.; Stollerman, J.; Rozenblyum, L.; Belfer, D.; Selim, A.; Mahoney, D.; and Steinbach, S. "A Telecommunications System to Manage Patients with Chronic Disease," in *MedInfo '98: Proceedings of the Ninth World Congress on Medical Informatics*, B. Cesnik, A. T. McCray, and J-R. Scherrer (eds.), IOS Press, 1998, pp. 1330-1334.

Glaser, B. G., and Strauss, A. L. *The Discovery of Grounded Theory: Strategies for Qualitative Research.* New York: Aldine, 1967.

Hunt, D. L.; Haynes, B.; Hayward, R. S. A.; Pim, A. M; and Horsman, J. "Automated Direct-from-patient Information Collection for Evidence-based Diabetes Care," in *Proceedings of the AMIA Fall Symposium*, D. R. Massys (ed.). Philadelphia: Hanley & Belfus, Inc., 1997, pp. 81-85.

Kaplan, B. "Addressing Organizational Issues into the Evaluation of Medical Systems," *Journal of the American Medical Informatics Association* (4), 1997, pp. 94-101.

Kaplan, B. and Duchon, D. "Combining Qualitative and Quantitative Methods in Information Systems Research: A Case Study," *MIS Quarterly* (12), 1988, pp. 571-586.

Kaplan, B., and Maxwell, J. "Qualitative Methods for Evaluating Computer Information Systems," in *Evaluating Health Care Information Systems: Methods and Applications*, J. G. Anderson, C. E. Aydin, and S. J. Jay (eds.). Thousand Oaks, CA: Sage Publications, 1994, pp. 45-68.

Kim, J.; Trace, D.; Meyers, K.; and Evens, M. "An Empirical Study of the Health Status Questionnaire System for Use in Patient-computer Interaction," in *Proceedings of the AMIA Fall Symposium*, D. R. Massys (ed.). Philadelphia: Hanley & Belfus, Inc., 1997, pp. 86-90.

Nelson, T. *Computer Lib/Dream Machines*, 1st ed. [n.p.], 1974.

Rae, J. B. *The American Automobile: A Brief History.* Chicago: The University of Chicago Press, 1965.

Shortliffe, E. H. "Dehumanization of Patient Care: Are Computers the Problem or the Solution?" *Journal of the American Medical Informatics Association* (1), 1994, pp. 76-78.

Strauss, A., and Corbin, J. *Basics of Qualitative Research: Grounded Theory Procedures and Techniques.* Newbury Park, CA: Sage Publications, 1990.

Turing, A. M. "Computing Machinery and Intelligence," *Mind* (59:236), 1950.

Turkle, S. "Computer as Rorschach," *Society* (17:2), 1980, pp. 15-24. Reprinted in *Managing Computers in Health Care: A Guide for Professionals*, J. A. Worthley (ed.). Ann Arbor, MI: AUPHA Press, 1982, pp. 223-243.

Turkle, S. *The Second Self: Computers and the Human Spirit.* New York: Simon and Schuster, 1984.

Turkle, S. *Life on the Screen: Identity in the Age of the Internet.* New York: Simon and Schuster, 1995.

Weiner, N. *God and Golem, Inc.: A Comment on Certain Points where Cybernetics Impinges on Religion.* Cambridge, MA: The MIT Press, 1964.

Weinzenbaum, J. *Computer Power and Human Reason: From Judgment to Calculation.* San Francisco: W. H. Freeman and Company, 1976.

Zuboff, S. *In the Age of the Smart Machine: The Future of Work and Power.* New York: Basic Books, 1988.

About the Authors

Bonnie Kaplan, Ph.D., is a Lecturer at the Yale School of Medicine's Center for Medical Informatics, a Senior Scientist at Boston University's Medical Information Systems Unit, and President of Kaplan Associates. A recognized expert in evaluating clinical applications of computer information systems, she is the author of over 30 refereed papers as well as numerous other articles and publications. She has worked with the U.S. Department of Veterans Affairs (VA) and the U.S. National Institute of Standards and Technology (NIST); the Universities of Pittsburgh, Cincinnati, and Chicago; and Yale, Boston, Harvard, and Johns Hopkins Universities. She is on the Informatics Advisory Board of the Pharmaceutical Research Institute of Bristol-Myers

Squibb, serves as Chair of the American Medical Informatics Association People and Organizational Issues Working Group, and as editor of the International Medical Informatics Association Working Group-13 (Organizational Impacts) newsletter. She has taught a variety of information systems courses in business administration and hospital administration programs at several universities. She holds a Ph.D. from the University of Chicago. E-mail: Bonnie.Kaplan@ yale.edu.

Ramesh Farzanfar received her Ph.D. in political science from Massachusetts Institute of Technology with a concentration in political communications and behavior and developmental politics. She has used ethnographic research methods in both her Master's thesis and her Ph.D. dissertation for which she interviewed 10 groups and 25 individuals respectively. She has taught at Mount Holyoke College and currently is a Research Associate/Systems Analyst at the Medical Information Systems Unit, Boston University Medical Center.

Robert H. Friedman, M.D., is an Associate Professor of Medicine (General Internal Medicine) and Public Health and Director of the Medical Information Systems Unit (MISU) at Boston University Medical Center. The MISU is a research and development units with a staff of eight clinicians, programmers and analysts and a number of trainees, working to create and evaluate computer systems that help physicians take care of patients. Dr. Friedman is also the director of the General Internal Medicine Fellowship Program at the Medical Center, a program that trains future general internal medicine faculty in research and teaching methods. Dr. Friedman is currently or has been the principal investigator on 22 externally funded research grants including 13 NIH grants. The research areas have included computer-based decision support, telecommunications, and medical education, with an emphasis on primary care, geriatrics, and cardiovascular disease and health behavior.

7 THE POTENTIAL OF THE LANGUAGE ACTION PERSPECTIVE IN ETHNOGRAPHIC ANALYSIS

Heinz K. Klein
Minh Q. Huynh
State University of New York, Binghamton
U.S.A.

Abstract

The purpose of this paper is to explore the potential value of applying language-based methods of analyses to textual ethnographic field records (like e-mail records) in addition to thick descriptions. Even though language analysis imports many a priori assumptions, they could provide more concise summaries than thick descriptions and add value to them. The methods of conversational analysis investigated here a based on a coding scheme derived from the Wittgenstein "language use" theory of meaning combined with Habermas' social action typology. The paper presents some coding issues for discussion that we encountered along with illustrative tabulations of sample coding results. Our purpose is to illustrate the potential value of language action coding to ethnographers both during the analysis and reporting stages of research.

Keywords: Ethnographic field study, thick descriptions, language action theory, coding schemes.

1. Introduction

The results of ethnographic field studies are typically presented as "hick descriptions" in the form of narratives or "stories." The disadvantage of thick descriptions is that they tend to be rather verbose and make it difficult to form a global picture of the social phenomena being researched. The general purpose of the research underlying this paper is to investigate the possibility of using some form of language-action analysis on the same textual ethnographic field records (such as tape recording transcripts or e-mail records) as are used to formulate thick descriptions. Such language-based analysis could yield more concise summaries than thick descriptions and thereby add value to them. We do not imply here that language action analysis can entirely replace thick descriptions, but at least they could help to identify "salient" or prominent characteristics in the ethnographic data. Hence, language action based forms of analysis would be secondary to thick descriptions, but they could play an important exploratory and summarizing role in ethnographic research. Our approach is rooted in the "semantic content analysis" in which coding is based on meaning rather than word counts. On the other hand, it is not a grounded theory approach because the researchers enter into the examination of the texts with an initial set of predetermined content variables extracted from our framework (Truex and Ngwenyama 1998, p. 452).

Specifically, we shall explore in this paper whether a descriptive analysis of the field records with a language action theory based coding scheme could shed additional light on the social phenomena to be studied. Examples of what we mean by social phenomena are the collaborative processes that occur in a group project for a college level course. Although group work in college is not the same as collaborative work in industry because of the "artificial" nature of the instructor centered classroom setting, student work groups should qualify as a convenient testing ground for new research methodologies. This paper will report some preliminary results of testing the application of the language action theory to obtain a global view of the nature of the social interaction in such collaborative student group settings. To avoid possible misinterpretation of the data analysis presented later in this paper, it is important to keep in mind that our purpose here is not to test any hypothesis or make any assertions on group processes. Rather, our intention merely is to explore the feasibility of drawing on the language action theory for the development of a coding scheme and illustrate its use for coding and interpreting a given "text." We seek feedback on the feasibility and potential pitfalls of our proposed methods.

The remainder of the paper is organized as follows. Section 2 explains the theoretical background for developing language action coding schemes and the setting in which the ethnographic "data," i.e., field records from participant observation and on-line group interaction transcripts, were collected. Section 3 presents some coding examples and reports the difficulties encountered in applying Habermas' action types to ethnographic "data." Section 4 analyzes tentative results from coding samples and section 5 summarizes our conclusions about the potential value added of using language action theory based coding schemes with ethnographic field studies.

2. Theoretical Background and Research Setting

Our long-term objective is to design a research methodology that will help us to understand the dynamics of group processes from an insider's perspective. As such an understanding is always incomplete, we intend to focus on such aspects as may be suggestive for designing technical systems that are compatible with social group processes. We presume that the understanding should come first and information technology design second.

We start with the common sense assumption that at the micro level all group interactions can be understood in terms of elementary acts or sequences of such acts. In some cases, these action sequences may be as simple as asking questions, accessing information, manipulating information, etc. In other cases, they may be quite complex, such as solving problems through sophisticated strategies, mitigating group conflict or building group cohesiveness and forming consensus through compromise and mutually accepted critique. In general, all these processes involve a series of transactions and transformations that lead to the creation, processing, distribution, use, control and management of data, information, and knowledge. The most obvious manifestation of the knowledge creation is in the ordinary speech and writings that are used as the medium of communication, including special symbols such as figures or models to support ordinary speech.

In accordance with the literature on speech act theory, language should be a form of human action. In order to capture the full meaning of language, we may need to resort to identifying the human actions that provide the ultimate rationale or purpose of one or several speech acts. Hence, we may need larger units of analysis than individual sentences or speech acts. Possible larger units are the social action types as Habermas (1984) proposed them in his "Theory of Communicative Action": instrumental, strategic, communicative, expressive, normatively regulated and discursive actions. Because these action types have already been widely discussed in the IS literature (e.g., Hirschheim, Klein, Lyytinen 1995 and 1996 or Cecez-Kecmanovic 1994), we will take them for granted here. However, a key point to keep in mind is that the action types are not easily identified at the micro level. Therefore, by themselves, they are not a good coding scheme. Nevertheless, the orientations underlying these action types can be very helpful to interpret the meaning of what people say or do and therefore helpful for constructing a more sophisticated coding scheme.

For the research site selected in this study, the following three orientations suggested themselves as the most important. The first is the instrumental (or means-ends) orientation to accomplish a set goal. The second is the communicative orientation, which prevails when people engage in communications to achieve agreement or shared understandings. The third orientation is called discursive and can be observed when conversation partners seek clarifications and reasons for claims made in the course of ongoing interactions. Hence, discursive action challenges what is being said, but seeks to restore agreement by giving good reasons abstaining from using force or other forms of power (e.g., court litigation). If the orientation is to get one's way (success) by means of power, this turns the communication partner into an opponent and the underlying orientation is instrumental (to achieve success in terms of one's own predefined objectives). This action type is called strategic, because it is directed against human opponents (or other organisms that can be expected to also act strategically) as opposed to inanimate objects, which don't have a mind of their own.

Speech acts are supposed to be the most basic units of ordinary speech, which at the vocal level consists of a series of utterances. Consistent with prior literature (i.e., Dietz and Widdershoven 1991) there are only four classes of speech acts, namely assertions, instructions, commitments (like promises), or declarations (like giving notice). As in prior research (cf. Klein and Truex 1996), we found it very difficult to classify all text that makes up ordinary group interactions in everyday life into one of those four speech acts. Hence we resorted to a typology that is richer and more flexible. Its original version was proposed by P. B. Andersen (1991). Based on Wittgenstein's analysis of language use, Andersen called the basic units in his typology "language games." These language games have different connotations from those in "mind games" or "games people play." The language games that seemed most appropriate for our research site are defined in Figure 1.

By analyzing what the group members tried to accomplish with their words in the group meetings, we sought to identify a set of language games to capture the universe of discourse of student group meetings, i.e., all that students typically say in study groups. We did this in two steps. First, a stable core set of language games could be transferred directly from prior work, i.e., Andersen (1991) and Klein and Truex (1996). As in prior research, the language games could easily be organized into a few fundamental types. Most likely these can be observed in many different contexts. Second, we identified additional language games that were not encountered in previous research. They are specific to the universe of discourse in the student group meetings studied here. Examples are the language games aimed at organizing the group or those aimed at accomplishing something specific such as tasks needed in writing a college report, the language games aimed at maintaining the group or smoothing social interactions or those aimed at mastering or controlling the technology. In general, the language games within each major type need to be adjusted to the specific research site context under consideration. Klein and Truex discuss the details of adjusting language games in developing a coding scheme.

The research setting for which the language games in Figure 1 were constructed consisted of carefully selected graduate-level courses with semester projects, because we did not have access to participant observation of an industry project group. The data used in this paper came from the "Leadership in Organizations" course. It is an MBA course offered to School of Management students who pursue the leadership certificate program. Although the field study at this specific site lasted only 15 weeks, its data interpretation benefitted from the background knowledge of a larger research project that involved several research sites and extended over a period of 18 months.

The major objective of this leadership course is for students to experience and gain insights into the emergence and development of leadership. There are several reasons why we believed that complex group interactions are likely to occur in the student project groups for this course and why this course would be a suitable testing ground for developing our research methodology. First, a unique feature of the teaching strategy in this course is the explicit role of the instructor as a facilitator rather than a knowledge disseminator. Students are expected to assume the ownership not only for their learning but also for that of others. Another important element of collaborative learning in this class is the emphasis on knowledge sharing. The course is carefully structured with tasks requiring students to share their work with each other.

Secondly, the course requirement of relevance to our research is the emphasis placed on many forms of student participation. These include in-class discussion, engagement in online chats, publishing work on the web and reviewing others' work. In addition, students are required to write a personal journal and an integrating essay to reflect their learning in the course. Finally, there is a group project that synthesizes the course materials covered into a full package of documentation and presentation.

A third attractive feature that makes this leadership course useful for research on group processes is the integration of user-friendly group support software into the course design. One of such groupware applications is TeamWave, which offers students a wide range of tools to use online. With TeamWave, students can work together at any time and in any place, e.g., chatting and brainstorming in a group, making a group decision or voting, arranging and organizing meetings, keeping track of and sharing the group's files, and so on. We took full advantage of the automatic transcript or recording facilities of TeamWave to obtain the best set of records possible.

The data presented in this paper was extracted from one of the case discussions in TeamWave. In this leadership class, the online case discussion was conducted regularly once a week outside of class time. At a scheduled time, students would log on TeamWave either locally from the school's public computing facility or remotely from their home or work place. The main objective of these online sessions was to discuss the reading materials or certain selective topics that the instructor assigned. In these online sessions, two specific tools of TeamWave—Brainstormer and Chat—were used to facilitate students' analysis of the assigned cases. In the Chat window, students had the ability to type their comments or to have a textual conference with each other. After typing in text and hitting the <Return> key, their statements would be displayed in the chat screen along with their log-in name. Everyone in the session would be able to see the statements from their screen. This is how users communicate to each other interactively online. Similar to the Chat, the Brainstormer also serves as a place for students to contribute their comments but in an anonymous mode because the students' names are not automatically displayed.

In principle, we proposed that understanding group learning in this leadership course involves analyzing the sequences of language actions. These language actions are performed by the actors under the influence of other components; namely, technology, pedagogical context, and group processes. It is hypothesized that all communicative transactions that occur in this context are driven by one of the three basic orientations that are at the core of Habermas' definitions social action types. The three orientations are instrumental, communicative, and discursive. They are associated with the group members' motivations and intentions when acting as group members. We assumed that one of the three orientations is dominant during one or more language games for each group member. This assumption led us to propose to combine the language games with Habermas' action types into the single coding scheme of Figure 2. In this coding scheme the orientations are coded as part of the five action types shown as column headers in Figure 2, i.e., the instrumental orientation is associated with instrumental and strategic actions while the communicative orientation is associated with normatively regulated and communicative actions.

3. Coding Issues

3.1 The Procedures Used in the Coding

From the online case discussions in the "**Leadership in Organizations**" course that we shall refer to as the H5 course, we selected two transcripts for the analysis in this paper. As mentioned earlier, these case discussions were conducted weekly in TeamWave and

Figure 1. Proposed Language Typology for the Analysis of Collaborative Group Processes (adopted and modified from Truex 1993, p. 100)

	Work Language Type	Characteristics
1. Language games establishing or changing the group work	1.1 Defining Tasks	activities aimed at determining the appropriate task and the order of performance of tasks
	1.2 Ordering	aims at allocating tasks to learners
	1.3 Work Distribution	aims at dividing a task or set of tasks among learners
	1.4 Work Coordination	aims at coordinating learners with the same tasks
	1.5 Work Priority	aims at changing priority so that one task preceded another
	1.6 Organization*	*When one learner asks another to take over a task for which he/she is better qualified and to designate a leadership role*
	1.7 Request for Suggestions and Assistance*	*aims at asking for technical helps, direction, and suggestions for possible ways of actions.*
	1.8 Reporting	*aims at informing persons about the current state of the work and responding to 1.7 above.*
	1.9 Monitoring and verifying*	*aims at controlling, monitoring, and verifying the quality, manner, and speed of the work in accordance to the requirements.*
	1.10 Requests for Clarification	used when assistance is needed to understand the task or interpret instructions or work methods
	1.11 Giving general advice and counsel	aims at giving knowledge about tasks or work organization and responding to 1.10 above. Tends to be repetitive and learning oriented

2.	Language games oriented to task completion	2.1 Problem Inception[+]	*activities aimed at problem identification, formation, and recognition; e.g., generating ideas and plans, asking personal perspective, raising opinions and evaluations, introducing issues, etc.*
		2.2 Data Interpretation	Activities in the task at hand devoted towards interpreting data, making it understandable, and converging ideas; e.g., giving ideas based on opinions and personal knowledge, expressing feeling, sharing personal perspectives on the issues, etc.
		2.3 Sub-goal Formation	activities aimed at partitioning a problem, decomposition of the goal or procedures used to accomplish a goal
		2.4 Problem Solving	*often refers to a one-time or non recurring situation. Attempts to clarify, synthesize, and reach a solution; e.g., reaching a consensus, deriving a correct answer, etc.*
		2.5 Execution[+]	activities aimed at performing a task, participating in a group's work, and interacting with other group members.
3.	Language games related to social relations, team building, and group norms	3.1 Accord	*Shows agreeing, concurring, accepting, complying, reaching consensus; e.g., negotiating and bargaining for consensus, sharing the same ideas or opinions, etc.*
		3.2 Disaccord[+]	*activities aimed at dealing with disagreements, different viewpoints, or conflicting interests within a group; e.g., rejecting an ideas, raising a challenging point, disagreeing with others' ideas, etc.*
		3.3 Greetings	serves as a signal for group bonding and for starting up and ending of a discussion.
		3.4 Collegial social conversation	serves to maintain communication open and build social relations. Often in the form of stories and analogies or jokes. A means of conflict avoidance and tension release; e.g., laughter, small talks, statement of forgiveness and uplifting the spirit, etc.
		3.5 Exclamations	outlets for emotions and often used to signal others about work progress, e.g., praise, supportive comments, etc.
		3.6 Grievance[+]	*Outlets for general complaints and discontent, e.g., confusing, worrying, overwork, time pressure, showing tensions, withdrawing, etc.*
4.	Language games associated with Technology and media	4.1 Structures imposed[+]	*Talks that are related to using the technology for task performance; e.g., scroll up the screen, vote in the voting tool, etc.*
		4.2 Media features[+]	*Terms that associate with the features of the tools, not necessarily related to the tasks; e.g., The PostIt, the chat works well.*
5.	Other unspecified games*		

Note: Significant modifications to the Truex's framework are specified in italic type and by the asterisk symbol "*". New additions of headings are indicated by the plus sign "+".

Figure 2. Proposed Data Coding and Analyzing Scheme for CSCL Systems

Work Language Type		FUNCTIONS				
		Production		Group Well-being and Support		Emancipation
		Instrumental	Strategic	Normatively-regulated	Communicative	Discursive
1. Language games establishing or changing the group work	1.1 Defining Tasks					
	1.2 Ordering					
	1.3 Work Distribution					
	1.4 Work Coordination					
	1.5 Work Priority					
	1.6 Organization					
	1.7 Verification					
	1.8 Monitor					
	1.9 Reporting					
	1.10 Requests for Clarification					
2. Language games oriented to task completion	2.1 Problem Inception					
	2.2 Data Interpretation					
	2.3 Sub-goal Formation					
	2.4 Problem Solving					
	2.5 Conflict Resolution					
	2.6 Execution					
3. Language games related to social relations, team building, and group norms	3.1 Seeking general advice and counsel					
	3.2 Collegial social conversation					
	3.3 Greetings					
	3.4 Acknowledging and reaffirming group bond and identity					
	3.5 Exclamations					
	3.6 Grievance					
4. Language games associated with Technology and media	4.1 Structures imposed					
	4.2 Media features					
5. Other unspecified games						

lasted about an hour. The average number of discussants was about 9-10 students. Because the discussions were text-based, all the communicative transactions were readily captured in transcripts as records. For simplicity, we limited our analysis on a sample of two sessions because our interest here is to explore the feasibility of the proposed coding scheme rather than to make any claims in group learning. In reading the transcripts from these meetings, we found sufficient evidence that they provide valuable insights into how a group works together on a complex task in a virtual environment over an extended period of time. However, we chose to present our analysis as illustrative examples of what other ideas can be gleaned from language action coding.

The first step in the coding was to convert the transcripts into text documents. Next, we imported these text documents into a qualitative data analysis program called ATLAS/ti. From the ATLAS/ti application, we began reviewing the transcripts. As we read and interpreted the statements in the transcripts (often with discussions among the authors), we assigned a code to a meaningful portion of text. One of the authors did the entire coding of the transcripts. Since we did not have outsiders available to serve as cross-coders and inter-coders, the other author had to be involved in the coding process. He checked the major sections of the coded transcripts from the first author. When incongruences arose, we met and resolved the differences ourselves. The codes were drawn from the predefined language types and Habermas' action types in the proposed coding scheme of Figure 2. The unit of analysis was based on one or more statements that expressed a complete thought or action; however, what constituted a complete thought or action was quite subjective. After coding all the transcripts, one of the authors ran queries from ATLAS/ti to obtain the frequency counts on each of the codes, i.e., selecting all the statements coded as "Act-Communicative". The results of these summary statistics will be presented in section 4. In this section we wish to focus on the difficulties that we encountered with resolving ambiguities as we applied the coding scheme.

3.2 Coding Difficulties

The sample transcript in Figure 3 is an excerpt from one online case discussion session in the H5 course (Transcript H50226, p. 2). Through TeamWave, learners could log in from different locations and participate in a real-time discussion. The Brainstormer tool in TeamWave was used to facilitate group interactions in this session. It supported a free format interaction that allowed learners to contribute their comments at any time. Another noticeable feature of this excerpt was its anonymity. Through the Brainstormer, learners could contribute either their comments or ideas without revealing their identity. As a result, it is not possible to trace the origin of these statements and associate them to a specific participant in these sample transcripts.

In this excerpt, the focus of the discussion was on the issue of leadership impact and the role of leaders in organizational change. In the contribution prior to line 68, one student "gave advice" "that there is a cycle of change that we need to pay attention to when effecting system-wide change." He then listed Nadler's five change activities: "(1) recognizing the change imperative, (2) developing a shared direction, (3) implementing change, (4) consolidating change, and (5) sustaining change." The language game of this passage was coded as "giving explanation" and the action type as

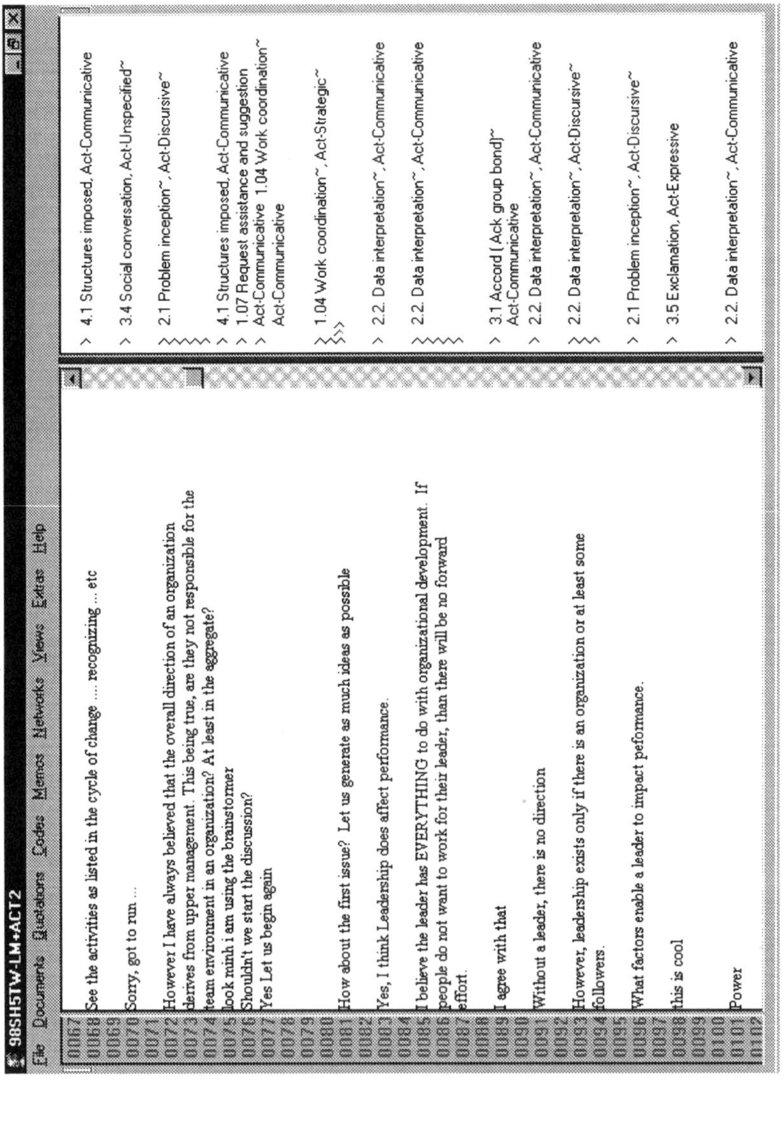

**Figure 3.   An ATLAS/ti Screen Capture Showing
a Sample Transcript with Codes**

"instrumental." After this, another student raised the following question ,"I wonder what the substantive and symbolic 'acts' would be with regard to this cycle of change. With the term 'acts,' do you mean the specific tasks of a team and leader or the overall goals?" This question was coded as "problem inception." The action type was coded as "communicative" and not "discursive" because the question did not really challenge any claims or assumptions from the preceding statements.

At this point, we will step the reader through a portion of the transcript and the coding shown in Figure 3. The result is shown in Figure 4. For brevity, we singled out only those statements from Figure 3 for discussion in Figure 4 that are ambiguous and problematic for the language action coding. From Figure 4, it will become apparent that the language game coding is relatively straightforward whereas the action type coding is much more ambiguous and subjective. Consequently, the reliability in the action type coding might turn out to be low.

The key point that we try to convey here is that the coding of action types relies on subjective judgments and contextual interpretation. As we shall describe in the analysis of coding summary, the difficulty is not the lack of the authors' experience or knowledge, but it is inherent in the whole approach of language action coding at its current state. There are simply no predefined rules to determine the proper unit of analysis. Should the social action types be analyzed at the sentence, paragraph, or issue-related level? In the next section, we will present our language action analysis based on each utterance. Although it is difficult to justify each action type coding, the result of cross coding both the language action types and the language games shows some interesting insights of collaborative group processes that might be hidden from other methods of analyses, i.e., thick descriptions, dictionary-based content analysis.

4. Analysis of Coding Summaries

As is evident from the prior discussion, we are faced with the following paradox. Whereas the coding of language games is fairly straightforward after some training and experience, it is insufficient to capture the full meaning of speech. Only if coding of language games is combined with action type coding can we capture some of the subtleties of the ongoing group interactions. However, coding for action types is full of ambiguities rendering the coding less reliable. This is true even for researchers who are very familiar with Habermas' Theory of Communicative Action. What seems to be surfacing here is some sort of linguistic Heisenberg principle: The more meaning we try to capture (i.e., the cross coding with language games type and action type), the less precise become the measurements (i.e., the ambiguities illustrated in Figure 4). The richness of cross coding can be elaborated in more detail with Figure 5.

Figure 5 was obtained by projecting the language games across different social actions of Habermas. The purpose of this cross tabulation is to explore the values and intentions underlying the language games. Our fundamental conjecture is that Habermas' five action types (namely instrumental, strategic, normatively regulated, communicative, and discursive action) represent most of significant social actions occurring in group processes. This study suggests that the values and intentions underlying different language games can be clarified substantially by cross-coding them with Habermas' action types. The reason for this is that the language games categories

Figure 4. The Illustration of Difficulties in Coding Language Action Types

Samples from the excerpt	Code assigned	Explanation
Line 68: See the activities as listed in the cycle of change…recognizing…etc.	Structured imposed	This refers to scrolling back to see things on the screen—a structure that exists as a result of using group supported technology.
	Act-communicative	The statement informs others for the purpose understanding.
Line 70: Sorry, got to run…	Social conversation	This does not relate to a task but it conveys a message so we coded it as social conversation.
	Act-Unspecified	The act here is unspecified. This is an example of ambiguity in coding social action types. The term "Sorry" itself might be treated as an expressive act, but taking the whole utterance into context, the statement could be interpreted as a communicative act. However, it is not clear in either case.
Line 72-74: However I have always believed that the overall direction of an organization derives from upper management. This being true, are they not responsible for the team environment in an organization? At least in the aggregate?	Problem inception	The language game is quite obvious in this case. The statement raised an issue for discussion. However, the language action type is debatable. Here, it was coded as discursive because the questions raised here put a spotlight on the common assumption. At the same time, this statement might also be interpreted as a call for an problem inception. If so, it would be communicative act.
	Act-discursive	
Line 81: How about the first issue? Let us generate as much ideas as possible.	Work coordination	The work coordination code seems reasonable. Again, interpreting this statement as a strategic act might require further elaboration. In this case, we speculated that this participant had a hidden motive when he/she made the statement. It could be that the individual wanted to shift the topic of the discussion and influence other to follow his suit. However, one might argue the communicative nature of the statement. The difficulty is how can we determine what is in the mind of the participants based on their speech expression.
	Act-strategic	

Result of coding Habermas action types in H5 case discussions. Note: X-axis: types of language move, Z-axis: action types, and Y-axis: percentage of statements

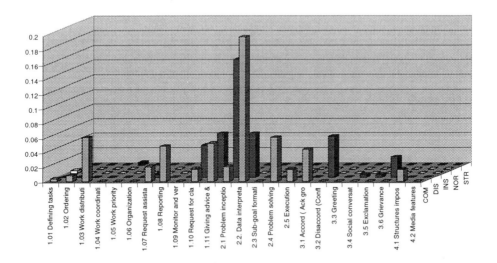

Figure 5. Result of Coding Language Games Cross Habermas' Action Types in the Case Discussions of the H5 Course

do not capture the underlying motives or intentions, which they serve. These motives are captured by the Habermas' action types. Hence language game coding alone is chronically insufficient whenever the same language game can serve different action types.

This observation is very evident from analyzing the following four language games in Figure 5: 1.10 Request for clarification, 1.11 Giving advice, 2.1 Problem inception, and 2.2 Data interpretation. These language games account about half of the conversation, hence are very critical communications. To clarify the point of chronic insufficiency of language game coding, let us single out 1.11, the "Giving advice" game. Clearly the meaning of "giving advice" is very different depending on whether the underlying orientation is to challenge some prior points (merely couched politely in terms of advice) or simply to offer some additional pieces of information to round out the picture that might have been overlooked by the listener. In the former case, the action type is discursive in the latter communicative. The giving advice game is almost evenly split among these two very different action types.

Similar observations can be made for the other three language game types. As shown in Figure 5, most of the problem inception language games are discursive, i.e., the interactions were aimed at challenging group members' opinions or positions. A similar pattern holds for the rRequests for data clarifications" (1.10). They also appear to be mostly discursive. The reverse is true for data interpretation language games: only one quarter is discursive and about three-quarters are communicative.

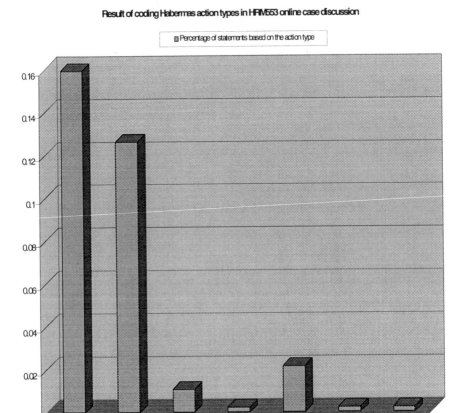

**Figure 6. The Result of Coding Habermas' Action Types in
a Sample of H5 Online Case Discussion**

In principle, we expect the distribution will vary depending on the social context in which the transcripts are generated, i.e., a consultation with the tax auditor would produce a distribution that is very different from that of a psychotherapy session. Given a very large number of texts and utterances, the shape of the utterance frequencies might not be necessarily stable because they are subject to the influence of various factors including group climate, task goal, group spirit, group composition, and so on.

There are a few more observations that can be made with the help of the coding summary in Figure 5. For instance, on the whole, the communicative action appears to be dominant (cf. Figure 6). This orientation indicates the values and intentions that the group members placed on achieving consensus and shared sense-making. Group members viewed themselves as partners engaging in negotiation and interpretation for shared understandings. This is further illustrated in the Figure 6.

Figure 6 shows the total number of statements in percentage with respect to different types of Habermas' social action types in all four online sessions of case discussion in the H5 course. From this figure, one interpretation of the group's interaction is that the group showed a high level of collaboration. It is evident in the large number of communicative actions that are oriented toward negotiation, coordination, and reaching consensus. In contrast, there are a smaller number of discursive statements that reflect disagreement, conflict, and argumentation. In hostile or confrontational groups, one would expect that strategic and discursive actions are much more frequent than were observed in our sample data. Hence the analysis method introduced could also provide of measure of the group climate on the cooperative vs. confrontational dimension.

In summary, adding the dimension of Habermas' action types reveals a deeper level of meaning in the group interactions. It shows us different orientations of the communication in which the group members engage. It allows us to take a closer look at the inside of group processes where values and intentions manifest themselves. It sheds light on the transformation where sense-making, shared meaning, and consensus building take place.

5. Conclusions and Further Research Implications

The paper reflects on the potential contribution from developing and using a coding scheme that is based on the language action theory. Based on an earlier study and its adaptation to a new setting, we concluded that such a coding scheme provides ethnographic researchers a valuable tool to examine group interactions at a deeper and more global level than the traditional thick description. Of course there are at least two challenges to this claim. The first is that the coding scheme rests on a number of a priori assumptions, which are imported by the researcher from "the outside" and therefore may not have any grounding in the field records. There is an obvious counter argument to this. Klein and Myers (1999) found that many ethnographic studies exhibit a heavy a priori theoretical bias even if they solely rely on thick descriptions because they tend to use external theories for interpreting the field records and for selecting the contents and format of their descriptions. Hence, no matter which research method is used, some theoretical bias is inevitable.

The second challenge that a language action coding scheme faces is the difficulty with achieving sufficiently reliable coding results. In other words, the coding methods need to improve the inter-raters' consistency and agreement. This is important for giving evidence for the plausibility and cogency of results. If coders trained in the same methods, but with otherwise different backgrounds, are presented with the same text, the tabulation of the results should be similar at least to the extent that they agree on the interpretation of the original thick descriptions. If the results vary with the coders, then they appear more or less arbitrary, because the reader cannot follow the logic of the coding in detail whereas thick descriptions are supposed to be sufficiently detailed so that the reader can form his own judgements.

However, even if coding consistency is not achievable, we made the case that the proposed language action coding scheme still is a valuable idea generation tool during the research process (the context of discovery). This is so because the coding scheme tabulation and interpretation are trustworthy only if supported by the thick descriptions. The latter are the warrant that the global characteristics captured in the tabulation of coding results depict qualities that are in some sense "true." True in this sense merely means that any claims made in the course of reporting results can be substantiated with the thick description accounts of the people's "lived experience," who perceived the phenomena about which academic claims are made. What is at stake here is the relationship between two socially constructed realities: those of the people whose lives are being studied and those of the researchers who try to understand them in their writings. We assume that the former is prior and hence any academic claims must be anchored to the lived experiences of the people whose experiences the academics try to understand.

References

Andersen, P. B. "A Semiotic Approach to Construction and Assessment of Computer Systems," in *Information Systems Research: Contemporary Approaches and Emergent Traditions*, H-E. Nissen, H. K. Klein and R. Hirschheim (eds.). Amsterdam: Elsevier (North Holland), 1991, pp. 465-514.

Cecez-Kecmanovic, D. "Business Process Redesign as the Reconstruction of a Communicative Space," in *Business Process Re-Engineering: Information Systems Opportunities and Challenges*, B. C. Glasson, I. T. Hawryszkiewycz, B. A. Underwood and R. A. Weber (eds.). Amsterdam: Elsevier Science (North-Holland), 1994.

Dietz, J. L. G., and Widdershoven, G. A. M. "Speech Acts or Communicative Action?" in *Proceedings of the Second European Conference on Computer Supported Cooperative Work: ECSCW'9*, L. Bannon, M. Robinson and K. Schmidt (eds.). Dordrecht, The Netherlands: Kluwer, 1991, pp. 235-248.

Habermas, J. *The Theory of Communicative Action*, Volume One, tr. T. McCarty. Boston: Beacon Press, 1984.

Hirschheim, R.; Klein, H. K.; and Lyytinen, K. *Conceptual and Philosophical Foundations of ISD*. Cambridge, England: Cambridge University Press, 1995.

Hirschheim, R.; Klein, H. K.; and Lyytinen, K. "Exploring the Intellectual Structures of Information Systems Development: A Social Action Theoretic Analysis," *Accounting, Management, & Information Technology* (6:1/2), 1996, pp. 1-64.

Klein, H. K., and Truex, D. P. "Discourse Analysis: An Approach to the Analysis of Organizational Emergence," in *The Semiotics of the Work Place*, B. Holmqvist, P. B. Andersen, H. K. Klein and R. Posner (eds.). Berlin: W. DeGruyter, 1996, pp. 227-268.

Klein, H. K., and Myers, M. D. "A Set of Principles for Conducting and Evaluating Interpretive Field Studies in Information Systems," *MIS Quarterly* (23:1), March 1999, pp. 67-93.

Truex, D. P. *Information Systems Development in the Emergent Organization*, Unpublished Doctoral Dissertation, Binghamton University, New York, 1993.

Truex, D. P., and Ngwenyama, O. K. "Unpacking the Ideology of Postindustrial Team-Based Management: Self-governing Teams as Structures of Control of IT Workers," in *Proceedings of the Work, Difference and Social Change: Two Decades after Braverman's Labor and Monopoly Capital Conference*, R. Baldoz, P. Godgrey, C. Jansen, C. Koeber, P. Kraft (eds.). Binghamton, New York, May 8-10, 1998, pp. 447-461.

About the Authors

Heinz K. Klein earned his Ph.D. at the University of Munich and was awarded an honorary doctorate by the University of Oulu, Finland, for his academic contributions to the IS faculty's research program. He is currently Associate Professor of Information Systems at the State University of New York, Binghamton. Well known for his contributions to the philosophy of IS research, foundations of IS theory and methodologies of information systems development, he has written articles on rationality concepts in ISD, the emancipatory ideal in ISD, principles of interpretive field research, alternative approaches to information systems development and their intellectual underpinnings. His articles have been published in the best journals in the field including *Communications of the ACM, MIS Quarterly, Information Systems Research, Information Systems Journal, Information Technology and People, Decision Sciences*, and others. He has also co-authored or edited several research monographs and conference proceedings in IS. He serves on the editorial boards of the *Information Systems Journal, Information, Technology and People* and the Wiley Series in Information Systems. He can be reached by e-mail at hkklein@binghamton.edu.

Minh Q. Huynh earned his Ph.D. in Information Systems and an MBA in the School of Management, State University of New York, Binghamton. He also holds a BA in Physics from Franklin and Marshall College, and BS in Computer Science from the University of Maryland. Prior to his academic career, Minh worked for in computing for four years with the U.S. federal government. His dissertation, *A Critical Study of Computer-Supported Collaborative Learning*, was completed in 1999. It combined analysis of ethnographic thick descriptions with focus group interviews and language analysis coding. His other research interests include application of social theory in IS research, the advancement of interpretive research and language action analysis methods. So far, he has published in the conference proceedings of the International Federation of Information Processing, International Resources Management Association, and North Eastern Decision Sciences Institute. He can be reached at br00328@binghamton.edu.

Part 2:

Field Studies

8 IDENTIFICATION OF NECESSARY FACTORS FOR SUCCESSFUL IMPLEMENTATION OF ERP SYSTEMS

A. N. Parr
Monash University
Australia

G. Shanks
University of Melbourne
Australia

P. Darke
Monash University
Australia

Abstract

The identification of factors which are necessary for successful implementation of enterprise resource planning (ERP) systems is of great importance to many organizations. ERP systems have to be configured and implemented, often by a team of business analysts and consultants over a period of months or years. The process is lengthy and expensive, and may include extensive business process re-engineering. Given that the investment in these systems, including both the package and associated implementation costs, is measured in millions of dollars, failure to meet deadlines and budgets may result in substantial company loss. However, the literature on the ERP implementation process, and the factors which either facilitate or impede its progress, is not extensive. This research reports the first stage of a research program which seeks to understand successful implementation of ERP systems. The objective, of the first phase was to identify what factors are necessary for successful ERP implementation, where success is understood as adherence to time and budgetary constraints. To accomplish this objective the authors studied 42 implementation projects by interviewing 10 senior members of multiple ERP implementation teams. Based on these interviews, 10 candidate necessary factors for successful implementation of ERP systems are identified. Of these 10, three are of paramount importance. They are management support of the project team and of the implementation process, a project team which has the appropriate balance of business and technical skills, and commitment to the change by all stakeholders. The next phase of the research will involve in-depth case studies to explore the relationship between these factors and broader contextual and process issues.

Keywords: Enterprise resource planning, success factors, system implementation, ERP implementation.

1. Introduction

One of the major information technology (IT) developments in the 1980s and 1990s has been the move by larger companies toward comprehensive, fully integrated software systems. Enterprise resource planning (ERP) is the generic name (Appleton 1997; Bowman 1997; Hecht 1997; McKie 1997) of this new class of packaged application software. The all-encompassing packaged solutions aim for total integration of all business processes and functions. A holistic picture of the business is represented in the information system within a single architecture. An ERP system has two important features. It facilitates a causal connection between a graphical model of key business processes and the software implementation of those processes, and it ensures a level of integration, and hence data integrity and security, which is not easily achievable with multiple software platforms.

The companies that create and deliver fully integrated software include SAP AG, Oracle Corporation, System Software Associates, Peoplesoft Inc., QAD Inc., Computer Associates International (CA) and Baan Co. The market leader is SAP, which has 39% of the world market (Piszczalski 1997). Software such as SAP's[1] R/2 and R/3 has offered companies a capacity to integrate their business plans and processes with their IT. The vendors of fully integrated software offer a package that is capable of processing all commercial functions of any company, no matter how large, diverse or geographically disparate the company's components may be. Moreover, the software is not limited to specific industry sectors: it can be configured for retail industries, mining companies, banks, airlines etc.

---

[1]Systems, Applications and Products (SAP) was developed by SAP AG.

A company's investment in ERP systems is measured in millions of dollars. A recent survey (Eckhouse 1998) of 200 IT executives in the United States found that 25% expected to spend more than $10 million on a complete implementation of their ERP system. In total, the cost of ERP investment in 1997 for 20,000 companies was $10 billion, which was an increase of 40% on ERP sales for 1996 (Martin 1998). The financial investment is both in the software package and in related services. These services include consulting, training and system integration (Caldwell 1998). In 1997, Gartner Group reported that "companies spent $19 billion" on ERP services (Caldwell 1998). The investment can also be measured in time. Implementation of the systems varies from six months to several years (Bancroft 1996; Bancroft, Seip and Sprengel 1998) and requires considerable infrastructure and planning. The size of implementation project teams varies with the scope of the project. For example, 300 people worked on GTE's 11 month implementation in the United States.

There is evidence (Ambrosio 1997; Fine 1995; Gartner Group 1998; Horwitt 1998; Martin 1998; Piszczalski 1997; Tebbe 1997) that both time and budgetary allocations for ERP implementation are being exceeded. Given this, plus the level of commercial investment in ERP systems, and the magnitude of the implementation task, it is critical to determine those factors which are necessary for an efficient implementation.

This paper reports the first stage of a research program which seeks to understand successful implementation of ERP systems. This stage is concerned with the identification of candidate necessary factors for ERP implementation success. We first review the general literature on system implementation and then on ERP implementation. From this, a list of candidate factors for successful implementation of ERP systems is synthesized. The factors identified in the literature are then used as one of the foundations for interviews with 10 senior ERP implementation managers. The other foundation is personal construct psychology (PCP). Because of the paucity of the literature on ERP system implementation success factors, this technique, which facilitates a less constrained response, is used to elicit factors not identified in the literature. The interviewees have been involved in 42 implementations of these systems and they equate "success" with the meeting of budgetary and time constraints. The interviews focussed on the factors which the practitioners believe, based on their experience, are the critical factors in determining that an ERP system will be implemented on time and on budget. The results of those interviews are then analysed and discussed. The next phase of the research will involve in-depth case studies to explore the relationship between these factors and broader contextual and process issues.

2. System Implementation Success

There have been numerous empirical studies on implementation success in information systems in general. Success has been described in terms of factors. Lists of factors can be misleading in that they ignore the relationship between the factors and organisational and cultural contexts (Bussen and Myers 1996). They also present a static perspective of success and do not capture the processes by which they operate and their interrelationships (Nandhakumar 1996). However, factors can be usefully combined with approaches which focus on understanding broader contextual and process issues to explain how and why factors and outcomes are related (Bussen and Myers 1996; Newman and Robey

1992). In the following section we review major studies on implementation success in order to infer candidate success factors for ERP implementation. First, we consider implementation of customized information systems, then implementation of packaged systems in general, and finally implementation of large, integrated packaged systems.

2.1 Successful Implementation of Information Systems

"Success factors" has been an influential conceptual construct for system evaluation in information systems and there is a large body of research (DeLone and McLean 1992; Rivard and Huff 1988; Swanson 1974) which uses this concept. As Delone and McLean note (p. 69 and Table 4, p. 72), user satisfaction is the most widely used single criterion of IS success, probably because it is relatively easy to measure. Ginzberg (1981) chose to view success as user satisfaction. This he measured in terms of actual use of the system and users avowed satisfaction with the system. Lucas (1979, 1981) also used user satisfaction as a measure of successful implementation. These studies were conducted in laboratory settings, or in response to survey questions. In the latter, some researchers (Bailey and Pearson 1983; Ives, Olson and Baroudi 1983; Raymond 1985; Swanson 1974) used multiple measures of user satisfaction.

Another alleged factor which leads to successful IS implementation is user participation in system design. In the 1960s, researchers considered user participation to be the key to the achievement of system quality, use and acceptance. Although the belief in the centrality of user participation in system design was strong, Ives and Olsen (1984) reviewed the relevant studies and found that strong evidence for its benefits had not been demonstrated. Nonetheless, in 1993 an issue of *Communications of the ACM* was devoted to participatory systems design. Barki and Hartwick (1994) distinguished

Table 1. Implementation Success Factors for
Information Systems in General

Factor	Study
Team skills/availability	Reich and Benbasat (1990)
Champion	Beath (1991), Kanter (1983), Rogers (1983)
User satisfaction	Bailey and Pearson (1983), Ginzberg (1981), Ives, Olson and Baroudi (1983), Lucas (1979, 1981), Raymond (1985), Swanson (1974)
User participation	Barki and Hartwick (1989), Keil and Erran (1995)
User acceptance	Benbasat and Dexter (1986), Davis, Bagozzi and Warshaw (1989), Ginzberg (1981), Robey (1979), Swanson (1987)
Usage	Baroudi, Olsen and Ives (1986), Lucas (1985)
Management support	Alter and Ginzberg (1978), Lawrence and Low (1993), Steinbart and Nath (1992)
Resources	Reich and Benbasat (1990)

user participation from user involvement and user attitude, and developed separate measures of each. Their belief was that earlier problems may have been due to lack of an adequate measure of the user participation construct. Keil and Erran (1995) provided the notion of "customer-developer links" to measure effective user participation.

User acceptance is another alleged factor which contributes to successful systems implementation. Performance gains from new systems are lacking when users refuse the new system either by not using it or using it badly. There have been numerous studies which seek to explain the key determinants of user acceptance (Benbasat and Dexter 1986; Davis, Bagozzi and Warshaw 1989; Ginzberg 1981; Robey 1979; Swanson 1987).

There are other success factors, which do not focus on the users but rather on the organizational context. Reich and Benbasat (1990) note that adequate resources and appropriate team skills are pivotal features of successful implementation. Also, the presence of a champion for the system has been studied widely. Beath (1991) claimed that the literature suggests that the presence and influence of a champion may even be "the most important antecedent of a successful implementation of a mission-critical system" (p. 355).

Champions are *not necessarily* people with overt power which derives from funds, resources and authority. Beath refers to these latter people as sponsors. Champions are usually managers who, by their zeal and foresight, overcome obstacles to successful implementation (Beath 1991). The main reason why these people are considered to be central to successful implementation is that they have skills that are critical for handling organizational change. These skills include enthusiasm, optimism, vision and a talent for conflict management (Beath 1991; Kanter 1983; Rogers 1983).

Another major success factor is the level of management support (Alter and Ginzberg 1978; Lawrence and Low 1993; Steinbart and Nath 1992). This factor is important for several reasons such as leadership of change management, the encourage-ment of support for the new system and handling of resistance to change. One would expect that this factor would be crucial in the implementation of ERP systems given the degree of reengineering.

2.2 Software Package Implementation

There is also a substantial literature on the implementation of packaged software (Gross and Ginzberg 1984; Lucas, Walton and Ginzberg 1988; Lynch 1984). Lynch reviewed his work with financial packaged software implementation and connected that experience with a Lucas' (1981) theory of software implementation. Noting that until then "implementation" and its "success" had been understood in the context of customized system development, he contended that implementation with packaged software had special meaning. He argued that the key differences are that those who implemented packaged software had a very limited influence on the technical quality of the system; that the ability to involve users is constrained in this context; and that there is less time to work on a relationship with clients. Lucas, Walton and Ginzberg described a model of the implementation of packaged software. Four sets of variables were hypothesized to be crucial to the success of the implementation process. These were variables concerned with the organization, the needs of the adopter of the software, characteristics of the package itself, and discrepancies between the capacities of the software and the company processes.

Additionally, there is a body of literature on large, integrated package implementation, particularly manufacturing resource planning (MRP) system implementation. MRP systems, like ERP, integrate a base system with other modules such as forecasting, general ledger, order entry and accounts payable and receivable, and are large systems which represent a considerable investment. Again, like ERP, MRP systems were reported (White et al. 1982) to have implementation failure rates that were high. They differ from ERP in that they are oriented toward the manufacturing industry only. Nonetheless, the literature on implementation of these systems is a strong source for candidate factors for implementation success with ERP systems. In the literature on determinants of success in MRP, both organizational and manufacturing factors have been studied. This paper concentrates on the organizational factors since those that are peculiar to the manufacturing industry are not relevant to ERP systems which are intended to be inclusive of all industry sectors.

Contributions on the organizational/behavioral factors that influence MRP include Greiner (1967), Greiner and Barnes (1976), Leavitt (1965), and Woodward (1965). Duchessi et al. (1988) conducted a nationwide survey in the United States to identify the steps that lead to successful implementation of MRP systems. This survey built on the work of White et al. and Landvater (1985). Some of the factors they surveyed are irrelevant in an ERP context (for example, the presence of a prior inventory system). However of those that are relevant, management support, appropriate training, and rigorous project management were confirmed by their study to be necessary for successful MRP implementation. A further study by Duchessi, Schaninger and Hobbs (1989) sought to describe the "critical factors which underpin a successful implementation" (p. 77). They considered three categories of success factors: commitment from top management, factors related to the implementation process such as training and project planning, and hardware and software issues. The last category is not relevant here, since this study is not concerned with package selection. Again, they concluded that top management commitment, which is demonstrated by, for example, establishment and membership of a steering committee, is essential for successful implementation. They also concluded that inadequate training may be responsible for unsuccessful implementation. A study by Ang, Sum and Yang (1994) also found that lack of comprehensive training programs hinders the MRP implementation process.

With respect to project planning, Duchessi, Schaninger and Hobbs (1989) also concluded that "successful and less successful companies were equally likely to use a formal project planning and control system" (p. 84). Therefore, factors related to project planning were not held to be determinants of MRP implementation.

As can be seen above, a large range of factors have been held to be responsible for successful implementation, and some studies have generated conflicting results. Moreover, the factors vary depending on whether we are referring to implementation within the context of customised system development or package implementation, and between small-scale, dedicated packages and large, integrated systems. These factors may or may not be relevant to ERP implementation success. In particular, the emphasis on the users does not at face level have any relevance. The studies on user participation, involvement and attitude concerned organizations that were developing new information systems; ERP systems are packages and users are not involved in their development.

ERP system implementation may be seen as more akin to project management than it is to system development. However, ERP systems are not simply installed. Over a period of months to years, users are heavily involved in the reengineering and configuration processes. Indeed, the project team is likely to be composed principally of users with business expertise (Bancroft 1996; Bancroft, Seip and Sprengel 1998), so their contribution is critical to successful implementation. On the other hand, some factors from this literature, such as management support and the presence of a champion, are plausible candidates, particularly in the context of substantial reengineering.

2.3 ERP Implementation Success

The literature that directly reflects on factors relevant for successful ERP implementation is not extensive. The systems are relatively new and a tradition of academic scrutiny and evaluation takes time to develop. What follows is a summary of this literature.

The studies by Bancroft (1996) and Bancroft, Seip and Sprengel (1998) provide nine "critical success factors" for ERP implementation. Many of these are consistent with the studies cited above. These include an emphasis on management support, the presence of a champion and an insistence on persistent communication with stake-holders. Factors that are important to successful project management, such as a good project methodology with clear milestones and appropriate training for the users and the project team, are also cited. However, some of the factors appear to be specific to ERP implementation. Businesses typically reengineer their businesses when they implement an ERP system. The level of reengineering implies that deep comprehension of business processes becomes a practical imperative for both process design and system configuration. The emphasis on reengineering leads to success factors such as the need to decide on substantial process changes prior to implementation; the need for a project manager whose skills range over technical, business and change management areas; and an understanding of the corporate culture which includes a deep analysis of the readiness and capacity for change. The nature of the implementation process also requires a project team composition contrary to that which might be expected in a normal large system implementation. Traditionally, the project team for such implementations is dominated by technical staff. With an ERP, the business analysts constitute 70% of the team and so it is important to choose a balanced team sufficiently flexible to adopt non-traditional roles. Finally, the length and complexity of the task require that teams should "expect problems to arise" and "commit to the change." This list of factors, which Bancroft provides, is derived from discussions with 20 practitioners and from studies of three multi-national corporation implementation projects. As such, it appears to be the most comprehensive and well evidenced of the guidelines which directly relate to ERP implementation, although the interview method is not described in detail. Bancroft, Seip and Sprengel noted that many of the factors are "classics" and not specific to ERP implementation. Nonetheless, they claim that each "takes on a greater significance" (p. 67) given the complexity of these projects.

Other researchers have views on successful implementation of ERP systems. Curran and Keller (1998, Chapter 3) argue that use of in-built templates, what they call the "blueprint" approach to reengineering is the key to fast implementation. Levin, Mateyaschuk and Stein (1998) contend that the empowerment of the project team and minimal customization are the keys to successful implementation. Martin (1998, p. 149)

Table 2. Common Success Factors

Factor	Information systems (general) implementation success factors	ERP systems (specifically) implementation success factors	Common factor?
Availability of skilled staff	Reich and Benbasat (1990)	Piszczalski (1997), Steen (1998)	Y
Champion	Beath (1991)	Bancroft (1996), Bancroft, Seip and Sprengel (1998)	Y
Management support	Duchessi et al. (1988), Duchessi, Schaninger and Hobbs (1989), Ginzberg (1981), Lawrence and Low (1993), Lucas, Walton and Ginzberg (1988), Martin (1998)	Bancroft (1996), Bancroft, Seip and Sprengel (1998)	Y
User satisfaction	Bailey and Pearson (1983), Ives, Olson and Baroudi (1983), Lucas (1979, 1981), Raymond (1985), Swanson (1974)		N
User participation	Barki and Hartwick (1994), Keil and Erran (1995), Lucas, Walton and Ginzberg (1988)		N
Project management	Duchessi et al. (1988)	Bancroft (1996), Bancroft, Seip and Sprengel (1998)	Y
Understanding of corporate culture		Bancroft (1996), Bancroft, Seip and Sprengel (1998), Ettlie (1998)	N
Process change completion first		Bancroft (1996), Bancroft, Seip and Sprengel (1998)	N
Communication		Bancroft (1996), Bancroft, Seip and Sprengel (1998)	N
Multi-skilled project manager		Bancroft (1996), Bancroft, Seip and Sprengel (1998)	N
Balanced team		Bancroft (1996), Bancroft, Seip and Sprengel (1998)	N
Methodology		Bancroft (1996), Bancroft, Seip and Sprengel (1998)	N
Appropriate training	Ang, Sum and Yang (1994), Duchessi et al. (1988), Duchessi, Schaninger and Hobbs (1989)	Bancroft (1996), Bancroft, Seip and Sprengel (1998)	N
Commitment to change		Bancroft (1996), Bancroft, Seip and Sprengel (1998)	N
Use of in-built ERP templates		Curran and Keller (1998)	N
Minimal customization		Levin, Mateyaschuk and Stein (1998)	N
Project team empowerment		Levin, Mateyaschuk and Stein (1998)	N
Use of standalone process modeling tools		Wreden (1998)	N

maintains that the most important success factor is "commitment to the process from the top levels of management." Langnau (1998) believes that the problem with implementation is under-estimation of budgets and time expenditures, so by implication the success factor is realistic budgetary and time targets and full testing. The pressure to implement quickly can, he suggests, result in inadequate testing. Ettlie (1998) suggests the key to success is "organizational culture change." His view is that the ability to manage major change is the key to successful implementation.

In the popular literature (*Information Week*, *Computerworld*, etc.), there are numerous "recipes" for success. Some (Caldwell 1998) see do-it-yourself (don't use ERP experts) as the key. Stedman (1998a) claims that a number of factors are impeding successful implementation. These include the level of reengineering required, a shortage of consulting professionals, lack of involvement by business users, the newness of the products, and the variation in set-up needs from site to site. Stedman (1998b) also suggests that unfriendly user interfaces and unforeseen costs in user training may be responsible for slow implementation. Steen (1998) suggests that a lack of experience among the IT professionals is a hindrance to successful implementation. Wreden (1998) contends that the use of process modeling tools is a factor in successful implementation. This contrasts with Curran and Keller, who claim the use of the ERP in-built tools is a preferred method.

2.4 Synthesis of Candidate Success Factors

If one synthesizes the literature from implementation of large systems in general and from implementation of ERP systems in particular, the list of success factors is long. Some alleged success factors are shown to be common to both (see Table 2); some are specific to information systems in general; and some are specific to ERP implementation.

In summary, the literature both on systems generally and ERP systems in particular revealed that the availability of skilled staff, the presence of a champion, appropriate training, rigorous project management and senior management support were all considered to be necessary. Many other factors (see Table 2) were identified but were not jointly cited. Because of the under-development of the research, all factors identified in both literatures became the basis for structured interviews. Further, the paucity of current research required the structured interviews to seek both validate factors identified to date and to elicit other potential factors.

3. The Study

Structured interviews were conducted with 10 implementation project managers, who between them had participated in 42 ERP implementation projects. Senior members of ERP project teams were interviewed in order to understand ERP systems implementation in practice and to elicit experienced practitioners' beliefs about factors which lead to successful implementation. The interviews were structured around the factors elicited from the literature review, but also provided scope for the interviewees to propose their own factors.

The method used to validate factors inferred from the literature was a structured questionnaire delivered in an interview format. The method used to elicit factors other than those derived from the literature was personal construct psychology (PCP). PCP was created by George Kelly in 1955 and later refined by Kelly and others (Adams-Webber 1979, Boose and Bradshaw 1987). This tool has been used extensively in knowledge acquisition research to model the cognitive processes of human experts, and so was thought to be appropriate for the elicitation of the concepts of success/failure from those who are expert in ERP implementation. PCP is a theory concerning individual and group psychological processes, which suggests that experts function as anticipatory systems, by which it is meant that they develop conceptual models so as to better understand and make predictions concerning their immediate world. At the center of the conceptual models are sets of dichotomous constructs, such as "soft/hard," "male/female," and "success/failure." PCP is well accepted in psychology, management and education, and is an underlying mechanism for knowledge acquisition techniques in much expert systems development (Gaines and Shaw 1988).

3.1 The Interviewees

The participants had been involved in a total of 42 ERP implementation projects in Australia and the United States. They consisted of project managers from within implementation companies and project managers and senior consultants from ERP consultancy companies. One participant had ERP implementation experience both in Australia and in the United States. On average, participants had been involved in 4.5 ERP implementations. The ERP systems included SAP R/2, SAP R/3, Peoplesoft 7, Peoplesoft 7.5, and Oracle.

The range of industries represented in the implementations by these participants was large and diverse. It should be noted that several of the very large companies had had several implementation projects, as they chose either a series of "little bang" or "phased" implementations. All companies were large and many were multi-nationals. So the expertise of the interviewees' represents "state of the art" knowledge of ERP systems implementation in a broad range of international companies and industry sectors.

A summary of the participants and their experience is provided in Table 3.

3.2 The Interviews

The interviews were conducted in person and a structured interview method was used. However, after the formal interview, a less formal interview to follow up on points of interest was frequently conducted. Prior to each interview, the interviewer collected data on the company and, on average, the formal part of the interview lasted one hour. At commencement of the interview, it was explained that the research focused on factors that facilitated the meeting of budgetary and time constraints. Many interviewees provided data such as presentations on ERP implementation given to their staff, post-implementation evaluation written reports, and documentation relating to project team membership, responsibilities and milestones. The interview questionnaire, notes, individual communications, and documentation served as the study database.

Table 3. Interviewees and their ERP Experience[2]

Current Position	Current Company Type	ERP	No. ERP Projects	Country	Industry Sector
Project Manager	ERP Consultancy	SAP R/2 R/3	8	Australia	Oil Chemicals Construction Automotive Medical Financial Services
Project Leader	Implementation company	SAP R/3	3	Australia	Chemicals
Project Manager/ Acct. Systems Manager	Implementation company	SAP R/3	5	Australia	Retail
Senior People-soft Consultant	ERP Consultancy	PS 7/7.5	4	Australia U.S.A.	IT Airline Medical
Principal People-soft Consultant	ERP Consultancy	PS 7/7.5 Oracle	7	Australia U.S.A	Oil IT Airline
Manager of H.R. and Payroll	Implementation company	PS 7/7.5 Oracle	7	Australia U.S.A	Public Sector Airline
Consultant/ Functional Leader Payroll	ERP Consultancy	PS 7/7.5	2	Australia U.S.A	IT Airline
Product Manager	ERP Consultancy	SAP R/3	4	Australia	Manufacturing IT Communications
National Manager - Systems Dev. IT	Implementation company	SAP R/3	1	Australia	Automotive
Technical Dev. Manager	Implementation company	SAP R/3	1	Australia	Medical
Total	**10**		**42**		

The structured interviews had several objectives. In Section A, the aim was to determine the experience of the interviewee. Depending on that experience, up to six implementations were then documented (Section B). The documentation included details of the ERP elements and the length and cost of implementation. Three different methods were then used to elicit success factors for ERP implementation. The first method (Section C) used PCP and the previously documented projects. Here the aim was to

[2]Note that *up to* six implementations per person were *documented*. Interviewees may have had more experience than this.

elicit, without prompting, the interviewee's beliefs about the factors which contributed to the success or failure of those projects. Three of the projects were selected at random, and then the interviewee was asked:

> *In what important ways are two of these projects the same, but different from the third, in terms of factors which you believe have contributed to the success or failure of the project?*

This procedure was repeated until no further success/failure attributes were forthcoming.

In Section D, the interviewees' responses were restricted to factors derived from the literature. This combination of factors was tested first by elicitation of a ranking on a Likert scale, then by elicitation of a Boolean scale. The ranking on the Likert scale ranged from 1 *(the factor is unimportant)* to 5 *(the factor is essential)*. The Boolean response *(the factor is/is not necessary for successful implementation)* aimed to overcome the tendency to name all factors elicited with the Likert scale as equally important, and thus to distinguish important from necessary factors.
So the question asked was:

> *If a factor is so important that, without it, an implementation could not meet time and budgetary constraints, please tick. (Otherwise leave blank.)*

The data gathered from "open" PCP elicitation and from the constrained responses to the literature factors was then analyzed to identify and delineate candidate necessary success factors for ERP implementation.

4. Analysis and Discussion of Results

This discussion is divided into the factors elicited from the less constrained sections of the interviews in which PCP elicitation was used and the factors gleaned from the response to those factors which emerged in the literature.

4.1 PCP Elicitation of Success Factors (Section C)

In this section, interviewees were directly reflecting on their experience of ERP implementation cases, and from these inferring factors which either facilitated or hindered successful implementation. As such, this section drew a more elaborate response from those with the most experience. All the factors were expressed either as "success factors" or as factors which had in their experience led to failure given that the factor was absent. So the presence of the factor is positive and its absence is seen as an inhibitor of successful implementation. Many factors were elicited, but factors which were unique or received little support are not reported here. Only one factor was elicited from all interviewees. Senior management support was held to be indispensable for successful implementation. Seven other factors were elicited from nine of the 10 interviewees. These eight factors are discussed below.

1. *Management support.* Management support was the one factor elicited from all interviewees. On the basis of those implementations which were successful in meeting budgetary and time constraints, all stated categorically that senior

management support was indispensable to the achievement of that success. Many stressed that such support was best when it was accompanied by close monitoring of and interaction with the project. This was demonstrated in several ways. On successful projects, management established a steering committee and one or more project teams. A member of senior management with responsibility for the implementation sat on the steering committee, and the project director had access to that person. Senior management delineated the functions of both the steering committee and the project Team(s), and established regular reporting mechanisms. Apart from these operational functions, senior management support was crucial to overcome the inevitable setbacks and conflicts that arise during such a large undertaking. A number of factors, including the level of business process reengineering, the depth and breadth of company change, conflict between consultants and in-house members of the project team, and the steep learning curve for project team members contributed to an atmosphere that was often stressful. The role of management in finding a way through the problems and tension was critical.

2. *The best people full-time on the project team.* Project teams typically consist of business analysts, users and technical experts from within the implementation company and consultants from external companies. Consultants regularly complained that the people released by the company were only intermittently available, and/or were not the most experienced and best available. In the words of one interviewee, the project team should have "the best people full time." The members of the company who are released to provide the business expertise, which forms the foundation for the new system configuration, should be the best available and should be released from all other duties to work on the implementation. This was not always achieved for several reasons. The "best" people are usually doing a core task elsewhere in the company, and releasing them raises problems of backfilling and its associated costs. Also, the people required need an overview of company processes, access to senior management, and to be multi-talented communicators with skills in technical, business, and people management. These people are critical to the whole implementation, and are involved in the design of the system, its testing and the training of the users. A related problem for the company is that such people are in high demand and the experience gained in the implementation may result in their movement out of the company. Several interviewees suggested that adequate remuneration was required to guard against this.

There was *one* exception to the view articulated above. One very experienced interviewee had developed a unique approach to team composition, which he referred to as a "virtual project team." These were employees who were scattered throughout the organization, were in constant electronic contact, and were dedicated on a part-time basis only to the project. All of these employees were selected on the basis of substantial company experience and individual talent. This method was not replicated elsewhere, although the interviewee had used it several times and believed it to be highly effective. It is worthy of further investigation.

3. *Empowered decision makers on project team and effective decision-making.* There were varying configurations of project teams: some divided the technical from the business experts; in some, the consultants were available only part-time; others placed all team members together in a "war-room" environment. However, most interviewees stressed that whatever the mechanics of the team, one factor was

crucial: the members of the project team(s) must be empowered to make quick decisions. Many reflected that delays in implementation had been due to slow decision making, because the decision had to be moved up the line, and/or wait upon senior management meetings. A related factor was the establishment of streamlined, publicly known decision-making processes. Determining and publishing the decision-making processes is an initial function of the steering committee.

4. *Deliverable dates* . Interviewees stressed that implementation was more likely to fail when the dates for deliverables were fluid and/or not communicated well in advance. There is one date that shapes all ERP implementation: the year 2000. Many of these systems are being implemented in order to achieve Y2K compliance for the company.

5. *Presence of a champion.* Although they did not distinguish champions from sponsors, interviewees agreed that the presence of a champion had facilitated many successful projects. If the person was a senior member of management so much the better, but often the person was the project manager. The role of advocate of the benefits of the system was crucial, particularly during the difficult times referred to above. This person was the one who was unswerving in promoting the benefits of the new system, even when users lauded (as they frequently did) the advantages of the old system.

6. *"Vanilla ERP": minimal customization and uncomplicated option selection.* One factor that related directly to the software was stressed. This was the need to choose minimal customization of the ERP. Where possible, the business should adopt the processes and options built into the ERP, rather than seek to modify the ERP to fit the business. The problem appeared to be that the belief that a business is unique is common, and leads to time consuming and lengthy customization. A related factor was selection of complicated ERP options during configuration when a simpler alternative was available. The practitioners know this approach of minimal customization and uncomplicated option selection as "Vanilla ERP." Two of the most experienced project managers claimed that this factor was in fact more important in achieving time and budget limits than any other factor.

7. *Smaller scope and functionality.* Not surprisingly, projects with smaller scope and functionality were likely to be more successful than more complex ones. The longest documented implementation took four years and was the most complex. It was a SAP R/2 implementation and all but one of the 12 main modules, and all associated submodules, were implemented. The shortest implementation documented took six months, but only two modules were implemented and the user group was relatively small.

8. *Definition of scope and goals, roles and responsibilities.* Successful system implementation means that the steering committee determines the scope of the project in advance and then adheres to it. One company changed the scope and resources four months into the implementation. Originally, they had intended to implement a SAP system in Australia then roll out the system across a range of Asian countries. Then came the South East Asian crisis and, four months into the implementation, the plan changed from a multi-national to a national implementation. This resulted in a change of consultancy company and a range of changes that affected both the timeline and the budget. This company expected the initial implementation to take 15 months and the final time was 27 months.

Table 4. Success Factors Elicited by PCP

Management	*Personnel*	*Software*	*Project*
Management support Deliverable dates communicated well in advance Empowered decision makers	"Best people full time" A champion	"Vanilla ERP"	Smaller scope and functionality Definition of scope and goals, roles and responsibilities

The above eight factors may be categorized as relevant to personnel, management, project management, or software. In terms of personnel, it was stressed that highly skilled business resources should be available full time on the project; that the presence of a champion facilitated success; and that there should be senior empowered personnel on the project team, so that rapid decision making was facilitated. In terms of software, there was substantial agreement that success arose from making as few changes as possible to the software and from selection of the most straightforward configuration options. In terms of the project itself, projects with smaller scope and functionality were likely to be more successful than more complex ones and some of the elements of all successful project management—for example, deliverable dates, good project planning, clear definition of goals, roles and responsibilities—were emphasized. Additionally, the importance of quick decision making and streamlined decision-making processes was emphasized. A summary of the findings of this part of the interview is provided in Table 4.

4.2 Response to Success Factors Reported in the Literature (Section D)

As previously stated, the factors which emerged from the literature were assessed first using a Likert scale and then a Boolean scale. The Likert scale ranked factors from 1 (unimportant) to 5 (essential). The only factor from the literature that was ranked, on average, by all participants as being very important to essential (4.5) was a balanced team. Five other factors were scored by the participants as very important. These were commitment to the change, communication, advocacy, corporate readiness and a multi-skilled project manager. Perhaps surprising, almost no one thought it was important to complete major process changes first. Also, it was considered by all except one participant that selection of a methodology was not particularly important. The list of average responses is shown in Table 5.

Some caution has to be exercised with these results. When all 10 interviewees' responses are tabulated, *all* factors in this section scored in the range *"important"* to *"essential."* However, when taken together with the Boolean *(necessary/not necessary)* responses, a sharper picture emerges. When encouraged to discriminate between those factors without which time and budgetary constraints could not be met, and those which, while important, may not be necessary, fewer factors were selected. However,

Table 5. Important Factors for ERP Success
(1 = unimportant, 5 = essential)

Factor	Average
Balanced team	4.5
Commitment to change	4.3
Communication	4.2
Advocacy	4.2
Corporate culture readiness	4.0
Multi-skilled project manager	4.0
User training	3.6
Methodology	3.5
Project team training	3.4
Complete business processes	3.1

Table 6. Necessary Success Factors

Necessary Factor?	Necessary	Not Necessary
Balanced team	8	2
Commitment to change	7	3
Communication	5	5
Project team training	4	6
Corporate culture readiness	3	7
Advocacy	3	7
Project manager	2	8
Methodology	2	8
User training	2	8
Completion of business processes	1	9

Table 7. Combined Success Factors

Factors elicited by PCP	Factors based on Response to Literature
1 Management support	1 Balanced team
2 Best people full-time	2 Commitment to change
3 Empowered decision makers	
4 Deliverable dates	
5 Champion	
6 Vanilla ERP	
7 Smaller scope	
8 Definition of scope and goals	

the ranking of the principal factors remained the same. Again, a balanced team was selected by eight of the 10 participants as being necessary, and seven said commitment to the change was necessary. These two factors were top scorers in the previous section.

All other factors were deemed necessary by *half or fewer of the participants*. The least necessary were completion of business processes (one interviewee), a multi-skilled project manager (two interviewees), a good methodology (two interviewees) and user training (two interviewees). The number of interviewees who selected a factor as necessary/not necessary is shown in Table 6.

If one combines the responses to the Likert and the Boolean scales, the literature factors which were validated are a balanced team and commitment to the change.

4.3 Synthesis of PCP and Literature Factors (Sections C and D)

If one combines the findings from the two sections, 10factors (see Table 7) emerge as candidate factors for successful ERP implementation.

Three of these factors are of paramount importance: these are management support, a balanced team, and commitment to the change. Management support is of paramount importance because it was the only factor elicited from all interviewees in the PCP section. A balanced team and commitment to the change were the only necessary factors to emerge from the literature response section. Additionally, the seven factors which were supported by nine of the 10 interviewees in the PCP section are shown in Table 7.

5. Conclusions and Further Research

This research has been concerned with the identification of ERP implementation success factors. Since it is a relatively new area, the approach has been to draw upon prior relevant research and then to synthesize that with the directly applicable research on ERP implementation success factors. Senior ERP practitioners were then interviewed using the results of the synthesized literature to guide the interviews. Of the 18 candidate factors from the literature (see Table 2, Common Success Factors), only six were confirmed by the interviews. These were management support, a champion, a balanced team, commitment to the change, minimal customization, and project team empowerment. The interviewees rejected 12 factors from the literature as being necessary for ERP implementation. Additionally, the interviewees proffered four factors that were not found in the literature. These are the best people full-time, smaller scope, deliverable dates, and the definition and adherence to scope and goals.

An analysis of the interviews demonstrates that there is a combination of success factors peculiar to ERP implementation. The three overriding factors are management support, a balanced team, and commitment to the change. A range of further factors (Table 7) has been identified as at least desirable. The prime implication of this research for practitioners is that the three overriding factors—management support, a balanced team, and commitment to the change—are *necessary* for successful implementation. Proceeding with an implementation when one or more of the factors is absent will lead to budget and time over-runs. The other seven factors are at least desirable, and so their absence should be grounds for concern.

Further research is necessary to refine the factors and to conduct in-depth case studies to explore the relationship between these factors and broader contextual and process issues. First, the three major factors in ERP implementation are in need of clarification. The interviewees have suggested elements of each. For example, it appears that establishment of a rapid decision making process is an element of management support. Similarly "management support" in this context appears to involve at least the establishment of a steering committee, close monitoring of the project team and the establishment of a clear public process for rapid decision making. Also, the emphasis on the best people to be released full-time to the project team raises a number of questions. How are the "best people" identified? What process ensures they are selected? The elements of each of these concepts are in need of further clarification. Apart from the three principal factors, the elements shown in Table 7 are clearly important, and it remains unclear whether they are necessary and/or desirable. Second, case studies can be used to explain the interrelationships between these factors, their interactions with organizational and cultural contexts, and the dynamics of the processes of ERP implementation.

The PCP methodology was found to be particularly useful in the elicitation of factors where existing literature is sparse. In this study, four of the 10 factors emerged from the PCP section of the interviews, rather than the existing literature. This technique, which draws directly upon the case experience of interviewees, provided wide-ranging and thoughtful data.

Another finding of this research is that much of the existing literature on system implementation does not have any direct bearing on ERP implementation. As stated, six of 17 candidate factors were validated by this research. Elements of traditional project management, an approach to the software, management support, and the personnel who are released to the project team combine to determine a successful implementation. The only factor from the existing general literature that is *unequivocally* validated by this research is management support. The literature on ERP system implementation success, scarce though it may be, is more accurate. Particularly, Bancroft's "nine factors" were all viewed as—at least—important. (However, when asked to identify factors that are necessary, only a balanced team and commitment to the change were viewed as necessary.)

The research in the existing general literature on the user's role (participation/usage/involvement, etc.) is not directly validated. However a "balanced team" is considered necessary, and it may be that a redefinition of the user's role would be worthy of further research. The elements of "a balanced team" were, to some extent, elucidated. A balanced team involved the right composition for the project team; a balance between users, business analysts, consultants and technicians. In the implementations documented, there was considerable variation in the team composition. The role of the users on the implementation team is an interesting new one, since they are not passive system users, but rather equals with the technical staff in that they contribute their business expertise. Case study research on these user experts, and on what constitutes, in this context, a balanced team, management support and commitment to the change, is required to provide a deep understanding of each of the factors and to strengthen and substantiate the findings of this study.

References

Adams-Webber, J. R. *Personal Construct Theory: Concepts and Applications.* Chichester, United Kingdom: Wiley, 1979.

Alter, S., and Ginzberg, M. "Managing Uncertainty in MIS Implementation," *Sloan Management Review* (20:1), 1978, pp. 23-31.

Ambrosio J. "Experienced SAP Users Share Ideas with Newbies," *Online News Story* August 27, 1997.

Ang, J.; Sum, C.; and Yang, K. "MRP 11 Company Profile and Implementation Problems: A Singapore Experience," *International Journal of Production Economics* (34), 1994, pp. 35-45.

Appleton, E. L. "How to Survive ERP," *Datamation* (43:3), March 1997, pp. 50-53

Bailey, J., and Pearson, S. "Development of a Tool for Measuring and Analyzing Computer User Satisfaction," *Management Science* (29:5), 1983, pp. 530-545.

Bancroft, N. *Implementing SAP R/3*. Greenwich, CT: Manning Publications, 1996.

Bancroft, N.; Seip, H.; and Sprengel, A. *Implementing SAP R/3*, 2nd ed., Greenwich, CT: Manning Publications, 1998.

Barki, H., and Hartwick, J. "Rethinking the Concept of User Involvement, *MIS Quarterly* (13:1), 1989, pp. 53-63.

Barki, H., and Hartwick, J. "Measuring User Participation, User Involvement, and User Attitude," *MIS Quarterly*, March 1994, pp. 59-79.

Baroudi, J.; Olson, M.; and Ives, B. "An Empirical Study of the Impact of User Involvement on System Usage and Information Satisfaction," *Communications of the ACM* (29:3), 1986, pp. 232-238.

Beath, C. "Supporting the Information Technology Champion," *MIS Quarterly* September 1991, pp. 355-371.

Benbasat, I., and Dexter, A. S. "An Investigation of the Effectiveness of Color and Graphical Presentation Under Varying Time Constraints," *MIS Quarterly* (10:1), March 1986, pp. 59-84.

Boose, J. H., and Bradshaw, J. M. "Expertise Transfer and Complex Problems: Using AQUINAS as a Knowledge Acquisition Workbench for Knowledge-based Systems," *International Journal of Man-Machine Studies* (26), 1987, pp. 3-28.

Bowman, I. "ERP 'Coned Off' Expect Delays," *Manufacturing-Computer-Solutions* (3:1), 1997, pp. 32-3.

Bussen, W., and Myers, M. "Executive Information Systems Failure: A New Zealand Case Study," *Journal of Information Technology* (12:2), 1996, pp. 145-153.

Caldwell, B. "GTE Goes Solo on SAP R/3," *Information Week*, June 8, 1998, p. 150

Curran, T., and Keller, G. *SAP R/3 Business Blueprint: Understanding the Business Process Reference Model*. Englewood Cliffs, NJ: Prentice Hall, 1998.

Davis, F. D.; Bagozzi, R. P.; and Warshaw, P. R. "User Acceptance of Computer Technology: A Comparison of Two Theoretical Models," *Management Science,* August 1989, pp. 982-1003.

DeLone, W. H., and McLean, E. R. "The Quest for the Dependent Variable," *Information Systems Research* (3:1), 1992, pp. 60-95.

Duchessi, P., Schaninger, C.; Hobbs, D.; and Pentak, L. "Determinants of Success in Implementing material Requirements Planning (MRP)," *Journal of Manufacturing and Operations Management* (1:3), 1998, pp. 263-304.

Duchessi, P.; Schaninger, C.; and Hobbs, D. "Implementing a Manufacturing Planning and Control Information System," *California Management Review*, Spring 1989, pp. 75-90.

Eckhouse, J. "Money Pours Into ERP," *Information Week,* June 15, 1998, p. 220.

Ettlie, J. "R-and-D and Global Manufacturing Performance," *Management Science* (44:1), 1998, pp. 1-11.

Fine, D. "Managing the Cost of Client-server," *InfoWorld* (17:11), 1995, p. 62.

Gaines, B., and Shaw, M. "Knowledge Acquisition Tools Based on Personal Construct Psychology," http://ksi.cpsc.ucalgary.ca/articles/KBS/KER/KER1.html#, 1988.

Gartner Group. "Implementing SAP R/3: Avoiding Becoming a Statistic," http://www.gartner.com/, 1998.

Ginzberg, M. J. "Early Diagnosis of MIS Implementation Failure: Promising Results and Unanswered Questions," *Management Science* (27:4), April 1981, pp. 459-478.

Greiner, L. M. "Patterns of Organizational Change," *Harvard Business Review* (45), 1967, pp. 119-130

Greiner, L. M., and Barnes, L. B. "Organizational Change and Development in Organizational Behavior and Administration," in P. R. Lawrence, L. Barnes, and J. Lorsch (eds.), Homewood, IL: Richard D. Irwin, 1976, pp. 625-636.

Gross, P., and Ginzberg, M. "Barriers to the Adoption of Application Software Packages," *Systems, Objectives, Solutions* (4), 1984, pp. 227-234.

Hecht, B. "Choose the Right ERP Software," *Datamation* (43:3), March 1997, pp. 50-53.

Horwitt, E. "Enduring a Global Rollout—and Living to Tell About It," *Computerworld* (32:14), April 6, 1998, pp. 8-12.

Ives, B., and Olson, M. H. "User Involvement and MIS Success: A Review of the Research," *Management Science,* (30:5), May 1984, pp. 586-603.

Ives, B.; Olson, M. H.; and Baroudi, J. J. "The Measurement of User Information Satisfaction," *Communications of the ACM* (26:10), 1983, pp. 785-793.

Kanter, R. M. *The Change Masters.* New York: Simon and Schuster, 1983.

Keil, M., and Erran, C. "Customer-developer Links in Software Development," *Communications of the ACM* (38:5), 1995, pp. 33-50.

Kelly, G. A. *The Psychology of Personal Constructs.* New York: W. W. Norton, 1955.

Landvater, D. *Control of the Business Newsletter.* Newbury, NH: The Oliver Wright Companies, 1985.

Langnau, L. "Glitz and Hype Signal Caution, " *Material Handling Engineering* (53:5), May 1998 **53:5**

Lawrence, M., and Low, G. "Exploring Individual User Satisfaction within User-led Development," *MIS Quarterly* (2), 1993, pp. 195-208.

Leavitt, H. J. "Applied Organizational Changes in Industry: Structural, Technological, and Heuristic Approaches," in *Handbook of Organizations,* J. G. March (ed.). New York: Rand McNally, 1965, pp. 1144-1170.

Levin, R.; Mateyaschuk, J.; and Stein, T. "Faster ERP Rollouts," *Information Week,* July 13, 1998.

Lucas, H. "Management Information Systems," *Management Accounting,* (61:6), 1979, pp. 8, 63.

Lucas, H. *Implementation, The Key to Successful Information Systems.* New York: Columbia, 1981.

Lucas, H.; Walton, E.; and Ginzberg, M. "Implementing Packaged Software," *MIS Quarterly,* December 1988, pp. 537-549.

Lynch, R. "Implementing Packaged Application Software: Hidden Costs and Challenges," *Systems, Objectives, Solutions* (4), 1984, pp. 227-234.

Martin, M. "An Electronics Firm Will Save Big Money by Replacing Six People with One and Lose All the Paperwork, Using Enterprise Resource Planning Software. But Not Every Company Has Been So Lucky," *Fortune* (137:2), 1998, pp. 149-151.

McKie, S. "Packaged Apps for the Masses," *DBMS.* (10:11), October 1997, pp. 64-66, 68.

Nandhakumar, J. "Design for Success? Critical Success Factors in Executive Information Systems Development," *European Journal of Information Systems* (5), 1996, pp. 62-72.

Newman, M., and Robey, D. "A Social Process Model of User-Analyst Relationships," *MIS Quarterly* (16:2), 1992, pp. 249-266.

Piszczalski, M. "Lessons Learned from Europe's SAP Users," *Production* (109:1), January 1997, pp. 54-56.

Reich, B. H., and Benbasat, I. "An Empirical Investigation of Factors influencing the Success of Customer-Oriented Strategic Systems," *Information Systems Research* (1:3), September 1990, 325-347.

Rivard, S., and and Huff, S. "User Developed Applications: Evaluation of Success from the DP Department Perspective," *MIS Quarterly* (8:1), 1984, pp. 39-50.

Robey, D. "User Attitudes and Management Information Use," *Academy of Management Journal,* September 1979, pp. 527-538.

Rogers, E. M. *The Diffusion of Innovations*, 3rd ed. New York: Free Press, 1983.

Stedman, C. "Business Application Rollouts Still Difficult," *Computerworld* (32:28, November 2, 1998a, pp. 53, 56.

Stedman, C. "ERP User Interfaces Drive Workers Nuts," *Computerworld* (32:44), November 2, 1998b, pp. 1, 24.

Steen, M. "Enterprise Resource Planning Teams Learn by Doing as Well as Training," *InfoWorld* (20:45), November 9, 1998.

Steinbart, P. J., and Nath, R. "Problems and Issues in the Management of International Data Communications Networks: The Experiences of American Companies," *MIS Quarterly* (16:1), 1992, pp. 55-76.

Swanson, E. B. "Management Information Systems: Appreciation and Involvement," *Management Science* (21:2), 1974, pp. 178-188.

Swanson, E. B. "Information Channel Disposition and Use," *Decision Sciences* (18:1), 1987, pp. 131-145.

Tebbe, M. "War Stories Outnumber Successes When It Comes to Implementing SAP," (19:27), July 7, 1997, p. 120.

White, E. M.; Anderson, J.; Schroeder, R., and Tupy S. "A Study of the MRP Implementation Process," *Journal of Operations Management*, 1982, pp. 145-154.

Woodward, J. *Industrial Organization: Theory and Parts*. London: Oxford University Press, 1965.

Wrenden, N. "Model Business Processes," *Information Week*, September 1998, pp. 1A-8A.

About the Authors

Anne Parr is a lecturer in the School of Business Systems at Monash University. She is currently completing a Ph.D. in ERP systems implementation at Monash University. Her research interests are in tourism and medical information systems, expert systems in medicine and ERP systems implementation. She has published research papers in several international conferences. Anne is the contact author for this paper and can be reached at anne.parr@infotech.monash.edu.au.

Graeme Shanks is an associate professor in the Department of Information Systems at The University of Melbourne. He holds a Ph.D. in information systems from Monash University. His research interests are in data warehousing, data quality, requirements modeling, and organizational implementation of information technology. He has published research papers in *Information Systems Journal*, *Journal of Strategic Information Systems*, *Information and Management*, *Requirements Engineering Journal*, *Australian Computer Journal*, *Australian Journal of Information Systems*, and *Data Warehousing Journal*.

Peta Darke is a senior lecturer in the School of Information Management and Systems at Monash University. She holds a Ph.D. in information systems from Monash University. Her research interests are in requirements definition, data quality, data warehousing, enterprise systems, and conceptual modeling. She has published research papers in *Information Systems Journal*, *Information and Management*, *Requirements Engineering Journal*, *Australian Computer Journal*, *Australian Journal of Information Systems* and *Data Warehousing Journal*.

9 HUNTING FOR THE TREASURE AT THE END OF THE RAINBOW: STANDARDIZING CORPORATE IT INFRASTRUCTURE

O. Hanseth
K. Braa
University of Oslo
Norway

Abstract

This paper tells the story of the definition and implementation of a corporate information infrastructure standard within Norsk Hydro. Standards are widely considered as the most basic features of information infrastructures—public as well as corporate. This view is expressed by a high level IT manager: The infrastructure shall be 100% standardized." Such standards are considered universal in the sense that there is just one standard for each area or function, and that separate standards should fit together: no redundancy and no inconsistency. Each standard is shared by every actor within its use domain, and it is equal to everybody. Our story illustrates that reality is different. The idea of the universal standard is an illusion just like the treasure at the end of the rainbow. Each time a standard, which is believed to be complete and coherent, is defined, the discovery during implementation is that there are elements lacking or incompletely specified while others have to be changed to make the standard work. This makes various implementations different and incompatible—just like arbitrary non-standard solutions. This fact is due to essential aspects of standardization and infrastructure building. The universal aspects disappear during implementation, just as the rainbow moves away from us as we try to catch it.

Keywords: Information infrastructures, standards, universals.

1. Introduction

Standards have usually been an issue related to phenomena being shared by large communities such as nations or even the whole world. Such standards are set by international committees and relate to issues ranging from measurement (the metric system) to telecommunications. As the number of information systems and the computing equipment grow inside organizations, the need to integrate them becomes crucial. Based on this fact, notions such as corporate IT infrastructures have gained attention (see, for instance, Weill and Broadbent 1998), and the definition and implementation of "corporate standards" have come into focus. As a company or corporation is different from a nation or the whole world, corporate standards might be seen as very different from "traditional" standards. We will show in this paper that they are not. Large organizations, through their globalization processes, are becoming too large and diversified for tight centralized control. At the same time, they are becoming increasingly embedded into different local environments because close customer contact is crucial for survival.

Corporate standards are almost non-existent as a research issue. Weill and Broadbent, for instance, simply declare that corporate infrastructures should be implemented by "defining and enforcing corporate standards." They do so without any discussion, which makes one assume that they think that it is self-evident that implementing standards is an important objective and that doing so is trivial. That is a serious misunderstanding.

With the emergence of various plans for making "National Information Infrastructures" it is widely accepted that standards are of crucial importance at the same time as existing strategies for their development and implementation are considered obsolete (Kahin and Abbate 1995). However, research to date seems not to have brought any pathbreaking results so far. No new approaches have been developed—beyond a consensus about learning from the Internet experience.

Standards are traditionally considered as purely technical and universal in the sense that there is *one* definition satisfying the needs for all users. This definition is assumed to be complete, ensuring that all correct local implementations will work in the same way. How to implement and use a standard is assumed given by the standard itself. This view is shared by engineers as well as managers, those involved in the definition of standards as well as their implementation. When implementing specific standards, it is commonly experienced that the assumptions do not hold. The problem, however, is then considered to be the specific standards themselves; they are incomplete and should be extended. For instance, when implementing standardized EDI solutions[1] between two (or more) organizations, the implementations of the standards are often found to be in conflict. This is seen as being caused by the incompleteness of the standards. One response to this has been the Open EDI standardization effort (Open EDI 1997). This effort is based on the assumption that the problems experienced will be solved by making the standards complete. "Completeness" will be achieved by standardizing not only the structure of the messages to be exchanged, but also their semantics and the

---

[1]EDI (Electronic Data Interchange) denotes the exchange of information between computers which traditionally are exchanged on various forms such as orders, invoices, customs declaration forms, etc.

organizational processes the messages will support. A similar approach is also chosen by the Open Distributed Processing standardization (ISO/IEC 1995).

2. Standards and Universals

This paper draws on research on issues related to standardization within the field of Science and Technology Studies (STS), trying to illustrate the fruitfulness of their findings for corporate as well as national or global information infrastructure standardization. The view on standards common among "standards designers," which we presented above, is not acknowledged in the STS field.

Communication protocols standards have not been much in focus within STS. On the other hand, standards, in a wide sense, are indeed *the* issue addressed in STS; in particular, standards in the form of universal scientific facts and theories and widely used technologies. These studies also, we believe, have something to tell us about information infrastructure standards.

Universality, actor network theorists have argued, is not a transcendent, a priori quality of a body of knowledge or a set of procedures. Rather, it is an acquired quality; it is the effect produced through binding heterogeneous elements together into a tightly coupled, widely extended network.

Perhaps the most basic finding within STS is the *local* and situated nature of all knowledge, including scientific knowledge. Latour and Woolgar (1986) describe how scientific results are obtained within specific local contexts and how the context is deleted as the results are constructed as universal. Universals in general (theories, facts, technologies) are constructed in this way: they become taken as given when the context disappears into a larger space of taken for granted assumptions. This construction process has its opposite in a deconstruction process when universals are found not to be true. In such cases, the universal is deconstructed by re-introducing its context to explain why it is not valid in the context at hand (Latour and Woolgar 1986).

In spite of the fact that the context of origin and the interests of its originators are "deleted" when universals are created, these elements are still embedded in the universals. They are shaped by their history and do not just objectively reflect some reality (scientific facts, theories) or exist as neutral tools (universal technologies). They embed social and political elements.

In the same way as other universals, infrastructure standards are in fact "local" (Bowker and Star 1994; Timmermans and Berg 1997). They are not pure technical artifacts, but complex heterogeneous actor-networks (Hanseth and Monteiro 1997; Star and Ruhleder 1996). When a classification and coding system such as the ICD (International Classification for Diseases) is used, it is embedded into local practices. The meaning of the codes "in use" depends on those practices (Bowker and Star 1994). The ICD classification system, developed and maintained by the WHO in order to enable a uniform registration of causes of death globally (to enable the generation of statistics for research and health care management), reflects its origin in the Western modern world. "Values, opinions, and rhetoric are frozen into codes" (Bowker and Star 1994, p. 187).

Berg and Timmermans argue that studies in the STS field tend to reject the whole notion of universals (Berg and Timmermans forthcoming; Timmermans and Berg 1997).

They disagree, saying that universals exist, but they are always embedded into local networks and infrastructures. They exist as *local universals*. They argue further that there are always multiplicities of universalities. Some of these will be in conflict. Each universal defines primarily an order it is meant to establish. Implicitly it defines at the same time a *dis*-order that does not match the standard. When a multiplicity of standards are involved in an area—which is always the case—one standard's order will be another's dis-order. Further, Berg and Timmermans show how a standard even contains, builds upon, and presupposes dis-order.

In parallel with showing how universals are constructed, STS studies have addressed more extensively how they are used, i.e., how they are made to work when applied in spite of the seemingly paradoxical fact that all knowledge is local. This is explained by describing how the construction of universals, the process of universalization, also has its opposite, the process of localization. The meaning of universals in specific situations, and within a specific field, is not given. It is, rather, something that has to be worked out, a problem to be solved in each situation and context. Working out the relations between the universal and the local setting is a matter of a challenging design issue. As a universal is used repeatedly within a field (a community of practice), a shared practice is established, within which the meaning and use of the universal is taken as given.

Just as the development of a universal is not a neutral activity, there are social and political issues involved in the use of universals. As their use is not given, "designing" (or "constructing") the use of universals is a social activity like any others taking place within a local context where social and political issues are involved.

We bring these theories to bear on the definition, implementation and use of corporate infrastructure standards. We will do this by first showing how the need for a universal solution—a standard—was constructed, how the decision to define and implement a corporate standard called Hydro Bridge subsequently was made, and the definition of its content. However, the main part of the article concentrates on the implementation of the standard. The most characteristic aspect of this implementation process is the repeated discovery of the incompleteness of the standard in spite of all efforts to extend it to solve this specific incompleteness problem. The process is a continuous enrolment of new actors and technological solutions to stabilize the network constituting the standard. This stabilization process never terminates due to the open nature of infrastructures, and because the standard creates disorder within exactly the domain for which it is designed and in which it is implemented in order to bring the domain into order.

The Bridge Standard in Norsk Hydro has been a success in the sense that it is widely diffused and several of its components are useful for large user groups. But it is a failure in the sense that there are still many users using products not complying to the standard and using the standard is highly problematic for larger user groups. The partial success is achieved because Hydro has been quite clever in doing all the required "real-time work" to "make the standard work." This work involves extensive efforts at adapting the standard to local environments, including developing and implementing quite a few gateways and converters and various ad hoc solutions, and accepting to live with inconsistencies and incompatibilities and managing them in a rather ad hoc fashion.

Implementing information systems has—just like standards—turned out to be nothing but a straightforward process. The problems encountered when implementing standards, seen as universals, are not just the same old problems repeated. The problems are new and different. To express it somewhat like a slogan, information systems gets implemented although they require continuous maintenance and other forms of "repair" work to keep it working, alive, on track, etc. A universal never gets implemented as such. The universal character disappears during implementation.

3. Norsk Hydro

Norsk Hydro was established in 1905. Fertilizer was the only business area until the 1950s, when Hydro started its expansion: moving into light metals and, later, oil and gas. These are the core business areas.

During the period 1972 to 1986, Hydro grew rapidly; its income raised from 1 billion to 60 billion NOK. The growth took place through acquisitions in agriculture (fertilizer) and light metals, and by building up brand new activities within oil and gas. Hydro's traditional style of management was to run the factories "hands off." This changed in 1985 when the decision was made to stop growing and concentrate on consolidating existing business.

Although a corporate IT department, called Hydro Data, existed, running the factories "hands off" implied independent IT strategies and solutions. Top corporate management acknowledged the view that IT was an important issue. Divisions had to work out IT/IS strategies, and they had to evaluate their own IT solutions' value for the company.

A central institution, called IS-Forum, was established and assumed responsibility for working out common strategies and policies concerning IT. A "Corporate Steering Group for IT" was set up as the legitimate unit for making decisions at the corporate level. Members of IS-Forum were primarily top IT managers in the divisions while the Steering Group was composed of high level managers in the divisions. The head of the Steering Group was one of the corporate vice presidents.

In 1990, a "consensus process" around common IT architecture started. Consensus was arrived at in 1991-92 about the importance of standards in general, the establishment of a shared TCP/IP based network, and the need for corporate standards concerning office automation/desktop applications. The corporate standard, defined a bit later, was named Hydro Bridge. Hydro Bridge is the standard on which we focus. During the 1990s the globalization process has gained momentum, and so has the focus on collaboration across all divisions and the development and use of a shared infrastructure.

4. The Hydro Bridge Standard: Constructing the Universal

We will now turn to the Hydro Bridge standard: its conception, definition, implementation and use. Bridge is seen within Hydro as the standard defining the infrastructure shared by the whole corporation.

The first PCs arrived at Hydro in 1983. PCs were introduced based on local initiatives. The oil community, being "allergic against mainframes" because they had to "turn around fast," was the first adopter. New PC technology was acquired as it appeared in the market: file servers and PC LANs, network operating systems, etc. The first Novell server was bought in 1987. The variety of products, applications, and configurations exploded, and the need for standardization was acknowledged. Standards for document templates, partition of disks, backup facilities and routines were defined and enforced. The pendulum moved back and forth between standardization and diversification for each new major "generation" of PC technology. The definition of the Bridge standard can be seen as a step in this process. It was now due time for standardizing PC desktop applications.

In 1992, poor integration among the IT mangers in IS-Forum was becoming widely acknowledged as a major obstacle for smooth operation of the company. There was a lack of integration and communication across divisions and between the divisions and the corporate headquarter, costs were too high, resource use was sub-optimal, etc. The most obvious answer to this problem was, for those concerned, standardization.

IS-Forum agreed with developing one corporate standard for desktop applications. As they knew others had a different view, the Hydro Bridge project was set up in November 1992 to look a bit deeper into the issue: analyzing costs and benefits, important obstacles, possible solutions, etc. The project was staffed by IT personnel. The leader was the IT manager in the oil refining and distribution division. The project soon proposed to the Corporate Steering Group was that a corporate standard should be defined and implemented.

Having decided that a standard should be settled, the next step was to define its content. It seemed obvious for all involved that this was a choice between Microsoft and Lotus products. Most saw Microsoft products as the clear winner. The Bridge project, however, decided, mainly due to costs, to go for the Lotus SmartSuite applications. The project members knew that this decision would be difficult to sell. To make that easier, they translated the issue about which producers' software to buy into a strategic one. To succeed in that effort, they allied with Lotus Notes: Lotus was chosen because Notes was a strategic tool!

Having decided on the content of the standard, there were still more issues to take care of. Among these were the scope, the reach and range, in Keen's (1991) terms, of the standard. Who should use it and in which functions or use areas? Initially, those advocating the Bridge standard meant that there should be no other systems used by anybody inside Hydro for functions which Lotus SmartSuite products covered. However, to obtain required acceptance for the decision, the Bridge project group had to agree to using Microsoft products in some areas. These included areas where large software applications were developed in Excel; for instance, applications for interpretation of data from lab equipment and for currency transformations in some budgeting support systems. Word was also accepted as the preferred word processor in several joint projects with other oil companies where the others required Word (or other Microsoft products) to be used as a shared platform.

Bridge as a corporate standard was formally approved by the Steering Group for Information Systems on April 29, 1994. But even in the very first version of the Bridge standard as defined at this moment, the ideal of having just one corporate standard was in fact already abandoned.

5. Product Development: Opening Pandora's Box

The step following the formal approval of the (first version of the) standard was its implementation into a "product." As the standard specified only a set of commercial products to be used, this might seem unnecessary. That was far from being the case. Products such as those involved here may be installed and configured in many different ways. To obtain the benefits in terms of less costly installation, maintenance and support of these products, they had to be installed coherently on all computers. Such a coherent installation is also crucial for establishing a transparent infrastructure where information may be exchanged smoothly between all users. Reaching these objectives, a considerable development task had to be carried out. The task included primarily bundling work in the form of developing scripts installing the applications in the same way "automatically." Developing these scripts was quite a challenge. Many unforeseen problems popped up, but the implementation of the first Bridge version of the desktop applications package was declared finished by January 1, 1995. When the product was launched, it was, however, far from being free of errors.

Until the product implementation project started, Bridge had been seen (or at least treated) as a self-contained package. During the product development, it was "discovered" that this was definitely not the case. To work as a shared infrastructure, this infrastructure itself required an extensive underlying and supporting infrastructure. However, the infrastructure underlying Bridge was far from standardized within the company. The major problems during the product development project were seen as caused by the lack of standardization of the underlying infrastructure. The implementation project tried solving the problems by standardizing each layer as they were uncovered.

The first and immediately underlying layer "discovered" was the operating system. Virtually all PCs were running DOS or Windows, so agreeing on Windows as the OS standard was not controversial.

A large number of PCs were running in local area networks running LAN software and possibly a networking operating system. This "layer" also had to be standardized for several reasons: most Bridge applications would be installed on a file server and not on each individual PC, the applications (users) were storing their files on such servers and using other shared resources such as printers, etc. On this level, a standard based on Novell's LAN products was specified. This included a design of a specific LAN topology to be implemented everywhere.

Dealing with the PC hardware layer was certainly the most demanding implementation challenge. PCs were discovered to differ significantly with respect to external device adapters (LAN, screen, keyboard, mouse, etc.) and their drivers, BIOS, memory, etc. This was so in spite of the fact that they were all "IBM standard" PCs. Later on, when laptops were included in the platform to be supported, PCMCIA cards created severe problems. Standardizing this layer would imply changing most of Hydro's PCs. This was obviously an impossible short term solution because of its costs. But it was considered a necessary long term solution—i.e., to be phased in over time.

The efforts aiming at strict standardization of PCs have been given up. It is simply beyond reach. The specification of the latest version of Bridge says that it will support any "standard" PC from any "major" manufacturers selling PCs "globally."

In parallel with the implementation of the Bridge infrastructure, communication generally has become more important. This implied that the global IP based network being built, Hydro InterLAN, also be included in Bridge. The underlying layers are indeed heterogeneous: two MBits leased lines, telephone services, broadband networks (ATM), radio and satellite communication (to oil platforms, for instance), etc. However, this heterogeneity has not caused any trouble since TCP/IP runs smoothly on top of all of them.

All major difficulties during product development were related to the desktop applications, not Notes. The desktop applications were brought into focus basically due to the fact that there was already an infrastructure—although a fragmented one—in place. The implementation of the Bridge standard implied that the existing infrastructure should be turned into a new one. In this process, several users (divisions) had to stop using their existing applications and switch to others. Concerning Notes, however, the infrastructure was designed from scratch. There was no infrastructure to be replaced: no *installed base* to fight (Arthur 1987; Hanseth forthcoming). Replacing one infrastructure with another is the most challenging issue, technologically as well as politically. Technologically because the transition will take some time, and during this period both the old and the new have to work and maybe even interoperate. Politically, simply because most users prefer the products and applications they are experienced in using.

The lack of standardization of an infrastructure's underlying layers will often be visible to those using and maintaining the infrastructure. This may imply that the infrastructure does not appear as unified and coherent, but rather as several separate and different ones. The Hydro Bridge standard was initially defined to save the expenditures on maintenance and support work. These expenditures were certainly lowered, but not as much as planned because several different Bridge implementations had to be maintained and supported on different platforms.

6. Diffusion, Adoption and Use: Meeting the Local

The common view seeing standards as universals means that the standard is just one thing equal for all. That is not how Bridge appeared as the adoption process unfolded. It was seen very differently by the different units due to differences concerning existing computing environment, available resources in terms of money and competence, cultures concerning management styles as well as use of technology, felt need for improved infrastructure, etc. The adoption speed and style also depended on the distance from the main office of Hydro Data. For those already using Lotus products, adopting Bridge meant doing almost nothing. Others had to change considerably. Bridge soon came to encompass several different systems. This implies that some implemented the whole package, others just a few components. In the latter group, you would find smaller offices in Africa, for instance, typically having just a few standalone PCs.

Strategies adopted for implementing Notes on the one hand and the rest of Bridge on the other have been very different. The desktop applications have been intensively pushed from the top. Notes, however, was initially not pushed at all. Later on, it was pushed as an e-mail system. Differences in strategies among the different units have implied that the Bridge is not implemented as one coherent universal package, but rather as many different ones which need to be integrated and linked together to make the overall infrastructure work.

The desktop part of Bridge diffused pretty fast. In April 1998, there were about 18,000 users, which means that it has diffused throughout most of the Hydro corporation. However, the diffusion speed and patterns have varied a lot among the divisions.

The Oil & Gas division is always the first to adopt any new technology, as it was with Bridge. However, Oil & Gas was a heavy user of Microsoft products, so the adoption of Lotus SmartSuite products has been somewhat mixed. The Lotus package is installed and used to some extent. Microsoft products were, however, still used heavily. This is partly due to local resistance among experienced Microsoft users, but, more importantly, the close collaboration with other oil companies using Microsoft products. The general rule established within this collaboration is that the company being the operator of an oil field to be developed determines what tools to use. That means that Lotus products are used in the projects where Hydro is the operator. In cases where a company using Microsoft products is the operator, Hydro has to use these products as well.

Within the oil sector, it has always been important to be advanced in using new technology to stay competitive. The actors in this sector have significant resources and most employees are highly educated engineers, always focused on finding better tools.

The large fertilizer divisions also adopted Bridge rather fast. The adoption was fairly smooth and easy as they already used Lotus applications, except the word processor. For them, adoption of Bridge basically meant switching from WordPerfect to AmiPro. They also used Novell's LAN technology already. Their existing network topology, however, was different from what Bridge specified. So, they had to restructure their network. This implied hard work for the technical personnel.

Other divisions were more reluctant because they had an installed base of solutions significantly different from what Bridge specified. In particular, adoption of Novell was challenging and expensive for those having large Banyan Vines installations. The transition to the Novell-based Bridge standard took time, and happened stepwise. Throughout this process, many different network structures were in operation. This required local customizations of other parts of the Bridge standard to make them run on top of the local networks.

The light metal (aluminum and magnesium) divisions have been slow in adopting Bridge. The aluminum divisions were fusioned into the company latest, having their own systems, which differed considerably from those found in most other divisions. In addition, they have a culture stressing local independence. For these reasons, they were negative toward Bridge and resistant to adoption. The magnesium division has also been slow in adopting Bridge. For them, the basic problem has been the costs. They are continually struggling with low income and have had problems finding space for Bridge investments in their budgets. This is in strong contrast to the Oil & Gas division.

There is also variation in diffusion speed within divisions. All new products and versions are first installed in Hydro Data, being the permanent pilot site. The next units are those physically located at Vækerø in Oslo, which is the largest Hydro Data site and the largest office in Norway housing major parts of the Oil & Gas and Technology & Projects divisions.

To make this heterogeneous infrastructure work, filters and converters for word processing formats still had to be used. In addition, different viewers are included in Bridge to let users easily get access to documents produced by tools they are not using themselves.

One of the desktop applications was more difficult to implement coherently in the organization than any other: e-mail. Companies always communicate with externals. With the diffusion of the Internet, supporting such communication by computers has gained much attention. Hydro has also adopted the Internet and integrated it with Bridge (discussed in a later section). However, they are already using several other computer networks for various purposes. They are developing a considerable network together with other oil producers and engineering companies working within the oil sector in Norway. They are using an X.400[2] based network as carrier for EDIFACT messages.

In finance and trading activities, they have been using Telex for a long period and they are, for instance, using a proprietary system delivered by Digital for communication with the aluminum exchange in London. They have even purchased a new computer-based Telex system, running on a PC under the CP/M operating system! These various e-mail and messaging systems are partly used separately, implying that quite a few users are using several of these systems. Others are integrated and interconnected through gateways. The new Telex system, for instance, is integrated into the overall message handling infrastructure. Telex messages can be sent and received as Notes e-mails through an X.400 system.

Hydro's policy, saying that they should use only Notes (and cc:mail up to now) for message based communication implies that these systems should be replaced. That has not happened, and there is no indication that it will, either. The use of most of these systems has deeply penetrated the work practices within which they are used, as most infrastructures do (Joerges 1988). And because Hydro is only one of many organizations included into these practices and networks, it is far beyond Hydro's power to replace Telex technology by other systems.

7. Applications Integration: Including the Environment

7.1 The Problem

We illustrated in section 5 how a smooth implementation of a standardized infrastructure on one level recursively requires a standardized infrastructure on the level below. A similar problem is found along the border between an infrastructure and its environment. When using the term context we are here referring only to *applications* which are not a part of the Bridge standard but related, in one form or another, to the applications included in Bridge. An application becomes a part of Bridge's environment when it is used within the same or related tasks.

There is an important difference concerning the relations between the Bridge standard/ infrastructure and its underlying infrastructure on the one hand and Bridge and other applications in its environment on the other. Bridge *requires* an underlying infrastructure, otherwise it will not work. Which applications populate its environment, however, is accidental. These are applications that the Bridge users decide to use in a

---

[2]X.400 is the ISO and ITU standard for e-mail.

way making them related or linked to each other. How these links are addressed and managed are accordingly different.

The applications that are included in another's environment can to some extent be specified at the same time as one is deciding to adopt an application. But the collection of applications used varies over time and so does how each application is used. Further, the way a user really uses her tools is to a large extent tacit. This means that an application's environment will be disclosed as the users go along using it.

Dealing with borders is also closely related to learning. As an infrastructure is used, one discovers new ways of using it, it drifts and "meets" other infrastructures. And this happened with the Bridge applications. One strategy for dealing with evolving use and drift is to include an application in the environment into the standard. We have mentioned above the inclusion of Microsoft applications and various "viewers." As the Internet was growing in popularity, its relationships to the Bridge applications became closer, leading to the situation where it was included in Bridge. Among others, a package of administrative applications (for smaller offices), called SUN, was included.

Some applications are linked together in a way making them interdependent. This leads to a need for standardizing the interfaces between them and how to use of them. Other applications are included because of relationships on a more abstract level. For instance, applications already included and some outside may be seen as "really of the same kind." So if one is included, then the others should be included as well.

The strategy followed in the examples above—i.e., when a relationship between one application inside the standard and one outside is discovered, the latter is included—might in some cases be the best one. But it does not work as a general strategy. Each component in a standard has its environment, and different components have different environments. Do the environments of different components need to be aligned? This means that environments are indefinite. Trying to solve this problem by extending the standard to cover what is linked to it will lead to indefinite regress. This means that one cannot solve this "border" problem. Further, borders cannot be drawn once and for all. They continuously have to be renegotiated and maintained through more or less ad hoc links and various forms of gateways.

We will illustrate in more detail the relationships between the inside and the outside and how changes on one side interfere with and affect what is on the other by looking at the integration and links between the Bridge infrastructure and SAP[3] implementations in Hydro.

7.1 SAP

The Bridge infrastructure has been built in parallel with a considerable SAP infrastructure.[4] These infrastructures were initially considered completely separate, but they have become increasingly intertwined as they have grown. The first SAP applications were

---

[3]AP is the leading product within the market for so-called Enterprise Resource Packages (ERP), which integrates applications for accounting, logistics, production control, etc.

[4]For a presentation and analysis of this SAP infrastructure, see Hanseth and Braa (1998).

installed in the agriculture division in France in 1990, and SAP was settled as the corporate standard in 1994. At that time, SAP implementation projects were going on in parallel projects in several divisions. As Bridge and SAP have been implemented in Hydro, they have also been closely tied together as illustrated by Figure 1. However, Hydro has never considered including SAP in Bridge. SAP is outside the scope of Bridge and it is a too big and complex an issue to be seen as just a part of Bridge. The relationships between SAP and Bridge have to be managed without being part of the same standard.

During the SAP implementation process, a wide range of links between SAP and Bridge were uncovered. Some of these were in fact "known" in advance and taken care of in the design process. This covers the part of Bridge that implemented infrastructural services required by SAP such as PCs, operating system, data communication network, etc. Other links emerged during the process. This includes services such as maintenance and support. Some important links still seem to be invisible to those involved. And the links do not always cause trouble. In some instances, SAP and Bridge are mutually dependent and are mutually enhancing each others' development and use.

Some divisions, for instance, discovered that the SAP applications have rather complex user interfaces. For infrequent users, this constitutes a big problem. Some divisions have tried to solve this problem by developing Notes interfaces to their SAP applications as well as others. This turned out to be quite a challenge, not the least because of SAP's policy in relation to allowing their customers to integrate SAP and other applications.

For some needs, data from SAP applications are extracted and made available through the Web-based intranet. Data are exchanged between SAP applications and others, in particular spreadsheet (1-2-3) and other Bridge applications. In some cases, data are transferred manually by means of cut and paste operations. In others, scripts and programs are developed to transfer data more or less automatically.

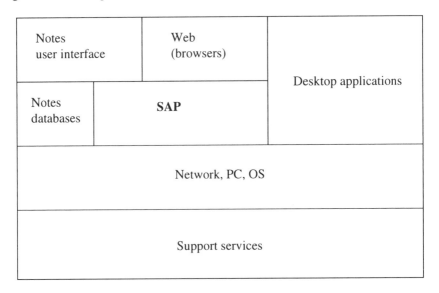

Figure 1. SAP's Embeddedness in the Bridge Infrastructure

When the Bridge standard was extended to included PCs, operating systems, and network protocols, this part of the standard was defined with a focus on the requirements of the Bridge applications. However, it was obvious that this part of Bridge also had to support other applications used by Bridge users, and accordingly it defined the infrastructure underlying SAP as well. Further, in some cases, SAP and Bridge had to rely on shared underlying services, for instance, user support. We will now look at a case from the European fertilizer division (called HAE) where SAP and Bridge required different kinds of services (as seen by those responsible for them).

The European fertilizer division has, since 1995, been developing a considerable SAP implementation to support a new organizational structure integrating all units in Europe. The SAP solution runs on top of the Hydro Bridge infrastructure as the SAP applications require, of course, PCs, operating systems, communication networks, etc. This part of Bridge turned out to fit SAP very well.

The implementation of Bridge in HAE turned out to be strongly influenced by SAP. Shortly after the decision to go for SAP, the IT manager concluded that Hydro itself did not have the resources and competence to take responsibility for the required data processing and operations services. HAE then decided to outsource these functions to a major global company offering such services.

The SAP transaction processing would run on computers physically located at a large processing center in the United Kingdom. When the decision about outsourcing SAP processing was taken, the IT management in the division thought that it would be an advantage if the same service provider also delivered the required network services connecting the client software on local PCs to the servers, so they decided to outsource that as well. Moreover, they also believed it would be beneficial to have just one provider responsible for the whole chain from the servers running the SAP databases through the network to the hardware equipment and software applications used locally. Accordingly, a contract was signed covering three areas, called processing, network, and (local) site management respectively. At this time, Bridge had been extended to include Hydro's global network. This contract meant than that the design and operation of the Bridge network was handed over to the service provider, as was the responsibility for installation and support of all elements of Bridge locally (PCs, operating system, desktop applications, the Notes infrastructure and applications, Internet software and access, etc.).

So far the outsourcing has been a mixed blessing. The network and processing services are fine, but site management (i.e., local support) has been problematic. The major problems seem to be related to the fact that the actual global service provider has organized its business in independent national subsidiaries, and is not able to carry out the required coordination across national borders. In addition, some problems are related to the fact that the site management contract specifies that users should call the help desk in the United Kingdom when they need support. The threshold for doing this is quite high for large user groups not speaking English, although the help desk should have people speaking all major European languages. When getting in contact with the help desk, problem solving is experienced to be much more difficult than when getting assistance from local support personnel. In this way, SAP has made the support of Bridge far more complex than desired. The site management contract was cancelled toward the end of 1998.

To make the SAP project succeed, people from all sites had to be involved to provide the project with the required knowledge about how tasks were performed and business was conducted at different sites. For a project of this size and distributed nature, smooth communication is mandatory. Notes applications have been used as the e-mail system, project document archives, and discussion databases. As such, Notes has been a crucial infrastructure, making possible the required cooperation between those involved all over Europe.

Notes has been widely used by virtually all SAP projects in Hydro, and SAP projects have in many divisions been the first users of Notes. In that way, SAP has been an important agent for making Notes diffuse. The initiatives for using Notes have been taken by IT personnel familiar with the technology and optimistic about its potential contributions to Hydro's overall productivity and efficiency. As all SAP projects are large and involve numbers of different user groups, knowledge about and practical experience with the technology become widely spread. SAP projects seem to be the most intensive users of Notes, and accordingly SAP one of the most important actors in making Notes diffuse in Hydro.

7.3   Notes and the Internet

There is often an overlap in functionality between components of an infrastructure and components in the environment. This raises the issue about which components should be used when and for which purpose. The relations between the Internet/Web and Notes are very much of this kind. Where to draw the border between the areas where each of them should be used is hard to specify. In almost any case, when the development of a Notes application is considered, one could just as well use Internet/Web technology. This means that where the border should be drawn has to be defined in every single case. Over time, an organization changes and technology evolves, implying that the border can drift (Ciorra 1996) significantly.

To avoid making the definition of the border a time consuming effort full of conflicts, a smooth interface is required. Inside Hydro, Internet technology is used for developing an "intranet." The Web technology overlaps a lot with Notes, so the Web-based intranet and the Notes infrastructure are integrated. In many cases, it has been considered a fairly open question whether to use Notes or Web technology. A Web interface to all Notes databases (through Domino) has been provided. The divisions developing Notes interfaces to their applications do the same for the Web.

As the integration between Notes applications and Internet (technology) was growing rapidly, Internet technology has also been put into the Bridge standard.

The close links between solutions based on Internet technology and Notes applications also include positive interference by causing a spill-over (Steinmueller 1995) in the sense that solutions available in the Internet world are also developed for Notes. For example, Hydro has also developed a search engine similar to those found on the Internet for searching across all Notes databases.

7.4 The Order's Dis-order

Standards and infrastructures interfere with each other. Sometimes a simple interface can easily be specified while the infrastructures stay rather independent. In other cases, they interfere in a way actively supporting each other so that each makes the other more useful. However, in some cases, they cause trouble for each other in a way that requires careful attention. The site management problems prove that what seemed to be a wise decision from a SAP point of view was at the same time a bad decision from the Bridge perspective. Standards are settled in order to create order. A smooth interaction between standards requires a global order. But such a global order is beyond reach in our complex world. We define local standards creating local order. Each local order interacts with others. And as long as there is no global order, one local order, however, well it is designed - will create disorder in its environment. SAP's order was Bridge's dis-order.

8. Maintenance, Support and User Training: Designing the Non-technological Elements of the Infrastructure

The Bridge infrastructure requires more supporting layers than operating systems, network services, etc. It also requires non-technical services: user training, maintenance and support (see Figure 2). Such services are equally as important as the technical layers such as operating systems and networks. The non-technical supporting infrastructure required by Bridge has been hard to establish. In fact, these are the required underlying infrastructures that Hydro has managed less successfully to implement. We will here mentioned three reasons for this.

First, these infrastructures are beyond the control of the Bridge team. This is illustrated by the SAP project presented above. For most divisions, the support services required by Bridge are seen as just a part of the overall IT support services within a division. How these services are set up, for instance, whether they are provided by an internal IT department, bought from Hydro Data, or outsourced to another organization, is—as in the SAP case—mainly based on what is believed to serve the "mission critical" applications best.

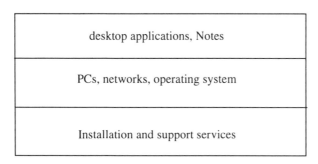

Figure 2. Applications and Required Underlying Services

Second, the problem is partly due to the fact that most of those involved in the design of the Bridge standard are technicians being blind to non-technical elements in the infrastructure. The problems related to lack of user support and training have only been addressed as far as they can see technological components as proper solutions. Advanced tools for IT infrastructure management, systems for "automatic" down-lowding of applications, a CD-ROM based training program, etc., have been developed and included in the Bridge package. However, very little seems to have been done to identify the needs for support and training, and how services satisfying these needs could be established.

The blindness to the non-technological elements is related to the third issue we will mention. There is huge local variation in the kinds of services needed as well as the bases upon which they can rely. The needs depend upon factors such as what kind of work is done, what kind of applications are used, which parts of Bridge are used, the general competence of the users, and, not the least, their knowledge about IT. How to establish support and training services depend, among other things, on what kind of resources—human and technical—are available. The kind of support services Hydro can offer its users in, say Africa, is rather different from what is easily available to those having an office in the same building as most Hydro Data people.

Hydro Data serves as the permanent site for pilot testing. As an IT department, the staff is very knowledgeable about how to use the Bridge tools. Further, the computing equipment and competence of the support staff are the best in Hydro. This is certainly far beyond what is found at smaller offices at remote locations. The competence level and services provided are taken for granted. The technicians do not see the role the support personnel are playing and how Vækerø differs form other sites in this respect. This fact is reflected in a statement expressed by users working in Hydro: "Bridge is the world as seen from Vækerø."

The technical and non-technical components of the supporting services are interdependent. Which non-technical (human, organizational) services are required depends on the design of the technical services. For instance, the need for user training will decrease if a carefully designed, computer-based user training package is provided. Further, the need for support also depends on how the technology is designed. Hydro Data has experienced, for instance, that equipment running at remote locations with limited support needs to be set up in a way that makes it more robust than otherwise. Disk drives are duplicated, more processing and storage capacity is provided, etc.

9. Standards Evolution: A Changing World

From its initial conception, the Hydro Bridge standard has changed considerable. Several new versions have been defined. Doing so is partly a result of learning and of Bridge's own success, and partly a result of the needs to adapt to a continuously changing world (also illustrated by the Internet example above).

Previous sections have mentioned many components being included in Bridge since its initial definition. We will mention some more: A second version of the Notes infrastructure was operational in May 1997. It introduced a new service providing high speed replication of databases following the established structure of hubs and spokes, a service providing replication directly between servers bypassing the hierarchical structure of hubs and spokes. Hydro templates for standard documents such as letters, memos, fax front pages, summons to meetings, minutes of meetings, etc., were defined, and a central directory service for resources across Hydro's different technologies is under development. The components included in Bridge are split into two categories.

Changing from one Bridge version to the next is a challenging task. Migrating to Bridge 97 means moving from Windows 3.1 to Windows 95 or NT. This implies that all applications had to be ported. This did not cause much trouble for most commercial products. But Hydro is using a wide range PC software developed in-house. Porting this software has been a major task.

The Bridge standard has been growing considerably since its initial conception, as illustrated in Figure 3 and Table 1. The character of the Bridge standard and infrastructure has also changed similarly. While it was clean and well structured at the time of conception, its evolution has caused an increasing number of parts which are overlapping and linked together in an increasingly more complex lattice, as illustrated in Figure 4.

Bridge has been growing along several dimensions: users and use areas, the number of applications, degree of duplication, and inclusion of the required underlying services.

In addition to the growth of Bridge from one version to another, the different speed of adoption and version updates among the divisions make the Bridge infrastructure a tremendous chaos.

Some divisions move fast to new versions, others are very slow. The file formats of desktop products change from one version to the next. The products are usually backward compatible so that newer versions can read files produced by older ones. The opposite is not the case. Accordingly, new product versions mean incompatibilities between tools used in different divisions. Bridge is growing from one version to the next, but a new version is only partially replacing the old. The old ones are still in use, which means that the new versions are introduced *in addition* to the old ones. All this means that the complexity, heterogeneity, and incompatibility of the infrastructure have been growing very fast.

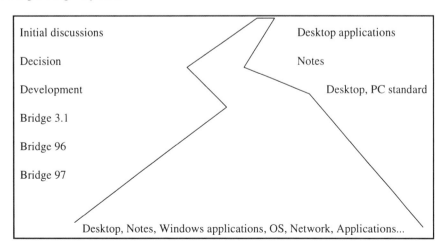

Figure 3. The Evolution of Bridge

Table 1. Bridge 97

Area	Products			
Information sharing	Notes	Web		
Desktop applications	Lotus	Windows		
E-mail	Notes	cc:mail	X.400	Telex
OS	Novell	Win95	WinNT	
PC	all major vendors			
Network	TCP/IP	Novell		
Telecom	Telephone	ATM	Radio	2MB
Support	Local	Hydro Data	Outsourced	
Versions	Bridge 3.1	Bridge 96	Bridge 97	

10. Conclusion

Standards are not universal—in the way usually assumed. They are only universal as abstract constructions. When they are implemented, they are linked to and integrated with local systems and practices, whether the applications are in the oil sector or telecommunication services (or rather the lack of) in Africa. The universality and homogeneity disappear as standards get implemented. They are locally embedded, in a sense making them part of the local, i.e., unique and non-universal. And they are continuously changing—in different directions in different localities.

Standards never creates order—in the way usually assumed. Order can only be created locally or as seen from one perspective. Dis-order is parasitic on order in the sense that creating order from one perspective means creating dis-order from another (Berg and Timmermans forthcoming). Managing dis-order—as dis-order—is just as important as creating order. One can never solve the dis-order problem by creating order. Doing that is just like trying to catch the treasure at the end of the rainbow. However, standards matter. The fact that standards are not universal does not mean that they are not important. They certainly are. Although infrastructures and standards get a local character as they are implemented and used, they do indeed also have some universals aspects. They are local and universal at the same time, i.e., they are *local universals* (Timmermans and Berg 1997). Standards are reducing the dis-order, but there will always be dis-order in terms of incompatibilities and redundancy. These issues have to be taken care of in terms of gateways, ad hoc patches, duplications, accepting to live with inconsistencies, etc. Successful implementations of infrastructures require skills in dealing with this just as much as skills in setting standards.

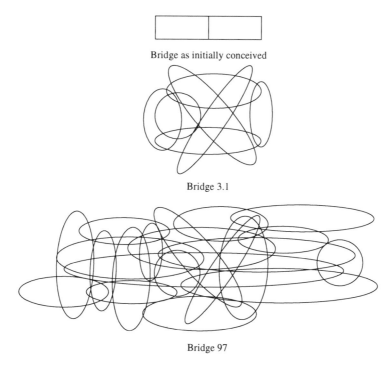

Bridge as initially conceived

Bridge 3.1

Bridge 97

Figure 4. The Evolution of Bridge

Acknowledgments

We are most grateful to Judith Gregory, Eric Monteiro, and Sundeep Sahay for their help in improving this paper.

References

Arthur, B. W. "Self-reinforcing Mechanisms in Economics," in *The Economy as an Evolving Complex System,* P. W. Anderson, K. J. Arrow and D. Pines (eds.). Reading, MA: Addison Wesley, 1987, pp. 9-31.

Berg, M., and Timmermans, S. "Orders and Their Disorders: On the Construction of Universalities in Medical Work.," forthcoming.

Bowker, G., and Star, S. L. "Knowledge and Infrastructure in International Information Management: Problems of Classification and Coding," in *Information Acumen: The Understanding and Use of Knowledge in Modern Business,* L. Bud-Frierman (ed.). London: Routledge, 1994, pp. 187–213.

Callon, M. "Techno–economic Networks and Irreversibility," in *A Sociology of Monsters: Essays on Power, Technology and Domination,* J. Law (ed.). London: Routledge, 1991, pp. 132–161.

Callon, M., and Latour, B. "Unscrewing the Big Leviathan: How Actors Macro-structure Reality and How Sociologists Help Them to Do So," in *Towards an Integration of Micro- and*

Macro-Sociologies, K. Knorr-Cetina and A. V. Cicourel (eds.). London: Routledge & Kegan Paul, 1981, pp. 259- 276.

Ciborra, C. U. "Introduction: What Does Groupware Mean for the Organizations Hosting It?" In *Groupware and Teamwork*, C. Ciborra (ed.). New York: John Wiley & Sons, 1996, pp. 1-19.

Hanseth, O. "Understanding Information Infrastructure Development: Installed Base Cultivation," forthcoming.

Hanseth, O., and Braa, K. "Technology as Traitor: SAP Infrastructures in Global Organizations," in *Proceedings of the Nineteenth Annual International Conference on Information Systems*, R. Hirschheim, M. Newman, and J. I. DeGross (eds.), Helsinki, Finland, December 13-16, 1998, pp. 188-196.

Hanseth, O., and Monteiro, E. "Information Infrastructure Development: Design Through Diffusion," in *First IFIP 8.6 Working Conference on Diffusion and Adaptation of Information Technology*, K. Kautz, J. Pries-Heje, T. J. Larsen, and P. Sørgaard (eds.), Norwegian Computing Center, Oslo, Norway, 1995, pp. 279-292.

ISO/IEC International Standard 10746-2 (ITU-T Rec. X.902): *Basic Reference Model of Open Distributed Processing—Part 2: Foundations.* 1995 (http://www.iso.ch:8000/RM-ODP/part2/toc.html).

Joerges, B. "Large Technical Systems: Concepts and Issues," in *The Development of Large Technical Systems*, R. Mayntz and T. P. Hughes (eds.) Frankfurt-am-Main: Campus Verlag, 1988, pp. 9-36.

Kahin, B., and Abbate, J. (eds). *Standards Policy for Information Infrastructure.* Cambridge, MA: MIT Press, 1995.

Keen, P. *Shaping the Future: Business Redesign through Information Technology.* Boston: Harvard Business School Press, 1981.

Latour, B., and Woolgar, S. *Laboratory Life: The Construction of Scientific Facts.* London: Sage, 1986.

Open EDI. *Information Technology: Open EDI Reference Model.* Draft International Standard. ISO/IEC 14662 (http://www.harbinger.com/resource/klaus/open-edi/model/oerm.html), 1997.

Star, S. L., and Ruhleder, K. "Steps Towards an Ecology of Infrastructure: Design and Access for Large Information Spaces," *Information Systems Research* (7:1), 1996, pp. 111-134.

Steinmueller, W. E. "Technology Infrastructure in Information Technology Industries," in *Technological Infrastructure Policy: An International Perspective,* M. Teubal, D. Foray, M. Justman and E. Zuscovitch (eds.). Amsterdam: Kluwer, 1995.

Timmermans, S., and Berg, M. "Standardization in Action: Achieving Universalism and Localization in Medical Protocols," *Social Studies of Science* (27), 1997, pp. 273-305.

Weill, P., and Broadbent, M. *Leveraging the New Infrastructure: How Market Leaders Capitalize on Information Technology.* Boston: Harvard Business School Press, 1998.

About the Authors

Ole Hanseth is an Associate Professor in the Department of Informatics, University of Oslo. His research focuses mainly on the interplay between social and technical issues in the development and use of large-scale networking applications. He can be reached via e-mail at Ole.Hanseth@ifi.uio.no.

Kristen Braa is an Associate Professor in the Department of Informatics, University of Oslo. Her work also focuses on the interplay between social and technical issues in the development and use of large-scale networking applications. She can be reached at kbraa@ifi.uio.no.

10 REENGINEERING THE SUPPLY CHAIN USING COLLABORATIVE TECHNOLOGY: OPPORTUNITIES AND BARRIERS TO CHANGE IN THE BUILDING AND CONSTRUCTION INDUSTRY

C. Sauer
K. Johnston
K. Karim
M. Marosszeky
P. Yetton
University of New South Wales
Australia

Abstract

Inter-organizational collaborative technologies provide potential competitive benefits based on time and cost advantages and value addition, especially when combined with supply chain reengineering. The building and construction industry would appear to be an ideal candidate for such IT-enabled reengineering because operations and project delivery are primarily organized around networks of collaborating organizations. This qualitative study of the Australian building and construction industry, however, finds a very low level of IT adoption. In explaining this phenomenon, we identify industry-level conditions as important factors influencing the low level of IT-based collaboration and suggest industry-level interventions which could stimulate both IT adoption and associated supply chain reengineering.

Keywords: Collaborative technology, supply chain, construction industry, information technology, barriers to adoption, reengineering.

1. Introduction

Inter-organizational collaboration, enabled by information technology (IT), is considered to be a source of competitive advantage for firms which reengineer their supply chain to optimize the benefits from collaboration (Chatfield and Yetton 1998; Johnston and Vitale 1988; Konsynski and McFarlan 1990). Advances in IT now allow firms to exchange information and share databases and business processes, with associated cost, time and value gains. Mutual business advantages to collaborating partners have been demonstrated in arrangements such as that between Wal-Mart and their suppliers (Janah 1998; Stalk, Evans and Schulman 1992). Much current economic and strategic commentary promotes a focus on core capabilities with outsourcing of all other activities (Quinn, Doorley and Paquette 1990). IT enables firms to outsource a range of operations to closely collaborating suppliers without losing control, because the technology makes the partners' businesses transparent to each other (Bensaou 1997; Snow, Miles and Coleman 1992). Even where there are not mutual benefits, power asymmetries have resulted in powerful corporations imposing inter-organizational collaboration through IT (Hart and Saunders 1997). Some proponents of collaboration encourage managers to believe that there is both a technical and strategic imperative driving them to transform their organizations into virtual corporations (Davidow and Malone 1992).

However, both adoption of and gains from IT-enabled collaboration are dependent to some extent upon factors such as the underlying exchange relationship, levels of interdependence, bargaining power and trust between the partners (Bensaou 1997; Choudhury 1997; Hart and Saunders 1997; Kumar and van Dissel 1996). Damsgaard (1999), for example, shows how an electronic market in freight handling in Hong Kong is unlikely to be fully developed because of the unequal distribution of benefits it would entail. More generally, some commentators have noted that game theory explains when collaboration is and is not likely to occur on the basis of the balance of benefits and costs for each party (Loebbecke, van Fenema and Powell 1999).

Most studies of IT-enabled inter-organizational collaboration have focused on specific industries (car industry, aircraft parts), typically characterised by limited types of inter-organizational relationships (powerful buyer, weak pool of suppliers) and using particular technologies (EDI). This paper examines an industry characterized by high levels of inter-organizational contracting, alliances and joint ventures—the building and construction industry—in which one would expect significant synergies from various forms of IT-based collaboration. However, very low levels of adoption of inter-organizational systems are found. The paper seeks to explain why this low level of IT use occurs and what actions may lead the industry to reengineer its supply chain to capture the benefits of IT-based collaboration. The paper identifies industry-level reasons for rationalizing the supply chain through IT and industry and organization-level factors that have militated against more than superficial adoption of collaborative technologies.

The contribution of this paper, therefore, is to demonstrate the importance of industry conditions in influencing levels of collaboration. In particular, it shows, notwithstanding the prevailing political wisdom of nonintervention by governments, that even if there are strategic benefits from collaboration, industry-level interventions such as government industry policy initiatives may sometimes be required to enable a collaborative dynamic. These findings are of particular interest because the Australian building and construction industry, which is the focus of this study, has for years operated to deliver projects through virtual or network organizations without inter-organizational IT.

2. The Australian Building And Construction Industry

The Australian building and construction industry undertakes all forms of building and construction including residential building, commercial building and civil engineering comprising houses, high-rises, offices, shopping centers, industrial plant and infrastructure. It accounts for 6.7% of GDP. This level of economic activity is achieved by some 150,000 companies employing 597,000 people. Average company size is only four people, with the majority being one person firms. The industry consists of several quite distinct sectors. These include the consultants principally involved in design work, such as architects, consulting engineers and quantity surveyors, and the contractors, comprising principal contractors who undertake actual building and construction, and specialist contractors (formerly referred to as subcontractors), who offer specialist building services such as concrete pouring, air conditioning provision and tunnel boring.

The industry structure is often described as "fragmented." Projects are typically delivered through a supply chain consisting of the consultants and contractors, coordinated sometimes by the principal contractor, sometimes by the architect, sometimes by the client, and sometimes by a combination of these. Each new project typically assembles a different set of consultants and contractors, often reflecting the developer's preferences rather than those of the consultants and contractors. Relationships among the different players are punctuated; they last for the period of a contract and may not be renewed for several years. There is understanding of each others' strengths and weaknesses but little or no trust. There is currently a perception among industry leaders that the supply chain is inefficient and requires reengineering.

Not only is the industry fragmented in terms of its different sectors, it is also highly localized. This is largely because of the location-sensitive nature of planning approval processes, although other factors such as geography and labor issues may also be relevant. The general trend to globalization has been slow to assert itself in this industry: a few globally known civil engineers do business in Australia; one major principal contractor, Concrete Constructions, is German-owned; a small minority of larger Australian companies have operations in Asia.

The industry has in the past suffered from a reputation for poor performance in relation to project delivery but the performance of the larger companies has improved significantly over the last 15 to 20 years. One of the causes of earlier problems with performance was poor industrial relations, but this has changed dramatically. Today, the industry is experiencing an Olympics-fueled boom. Despite high demand, the industry continues to be highly competitive with many companies experiencing very low margins. It is agreed throughout the industry that business is cost-driven. The emphasis on cost comes from the customer, often property developers whose principal interest is their margin in on-selling a project. Issues such as building performance and maintenance profile over its lifetime are relatively unimportant to the immediate customer and so are often sacrificed to cost.

One outcome of the industry's fragmentation and low margins is a continuing power struggle between the different sectors. This is most visible between architects and principal contractors. Architects are keen to regain control of the building process while principal contractors increasingly seek to undertake "design and construct" work. More generally, each sector is aware that if it could break into other sectors, not only could it appropriate their profits, but it could also streamline coordination between their activities. This situation is complicated by coopetition (Loebbecke, van Fenema and Powell 1999), that is firms may compete for a contract and then cooperate as contractor-subcontractor on that contract, and they may also cooperate as business partners in subsequent tenders.

3. Study Method

The objective of this study was to understand how IT can provide long-term benefits to the building and construction industry through inter-organizational collaboration across the supply chain. As there have been relatively few studies of the strategic application of IT in building and construction (Ahmad, Russell and Abou-Zeid 1995; Brandon, Betts and Wamelink 1998), with the exception of technical research relating to computer-assisted architecture, we saw ourselves as undertaking exploratory research and therefore adopted a qualitative approach.

The fieldwork was undertaken by the authors between June and August 1998. In order to achieve broad coverage, we segmented the industry into sectors (architects, quantity surveyors, engineering consultants, contractors [principal and specialist] and building manufacturers and suppliers). We divided companies by size from large to small-to-medium-sized enterprises (SMEs). We selected firms to study on the basis of a matrix of sectors and sizes, and added some major clients. In total, we interviewed one or more senior managers from 30 organizations using a semi-structured interview format based on an initial protocol of common questions.

The interviews were conducted primarily in firms with a presence in the Sydney region, although many also operated in other Australian states. We visited companies located in outback Australia, and conducted telephone interviews with companies as far afield as Perth. In order to reduce the probability of bias, we interviewed senior managers of acknowledged technology leaders as well as less advanced technology users in all industry sectors. We sought to identify actual and potential business benefits from IT-enabled collaboration and barriers to, or inhibitors of, adoption. Most interviews were conducted by at least two members of the study team. All were recorded, written up within 24 hours, and copied to the other members of the team.

Both during the fieldwork and on its completion, we sought to identify variant forms of supply chain and the different ways companies could add value within them. We listed the barriers to IT-enabled collaboration, then separated them into those that could be addressed at the industry level and those that could not. This formed the basis for devising recommendations for policy and practical steps to encourage successful reengineering of the supply chain.

4. Findings

4.1 The Supply Chain Opportunity

All the leading industry thinkers whom we interviewed had identified the need for change. They identified four underlying reasons:

1. Threat of global competition. In an era of globalization, any firm anywhere in the world that can leverage IT to enable it to compete differently will be a threat to all others in the industry.

2. Underutilization of IT. Building and construction is an information intensive industry that has as yet used IT for little more than personal productivity benefits.

3. Supply-chain inefficiency. It is recognized that the current supply chain is characterized by concentration of specialist activity within a specific stage, but with almost no sharing across organizations across stages. There is little understanding of prior decisions and little preparedness to combine different knowledge. For example, architects often ignore the issue of how "buildable" their designs are for a contractor and, equally, contractors may compromise a design feature during building because they do not understand the design rationale.

4. Availability of enabling technology. Various software technologies are available to permit more intensive cross-sectoral collaboration.

We found that the source of the greatest potential for transforming the industry, beyond another round of driving down costs, lies in reengineering the supply chain to deliver increased value for the client. There are three levels of potential benefit from cross-sectoral collaboration:

Level 1 IT can be, and typically is, used to improve the efficiency, speed and quality of communication across sectors, thereby reducing cycle times and making a small gain in quality for the whole supply chain.

Level 2 IT can be, but typically is not as yet, used to facilitate the creation of a transformed supply chain. By taking a different approach to cross-sectoral relationships, for example by encouraging greater concurrence between tasks conducted by firms in different sectors through greater sharing of information, it may be possible to achieve substantial savings in time and money for the client.

Level 3 In a supply chain characterized by the sharing of information and knowledge, the potential exists to increase the total value to the ultimate client (the developer/operator of the building or plant) by improving performance on multiple dimensions. For example, if architects, engineers, contractors and clients use appropriately rich communication channels to share information when a design is first being conceived, it will be possible to design and build more efficiently, with less difficulty, and with far greater benefit to the customer. New kinds of solution developed collaboratively would potentially be safe, aesthetic, easy to build, provide better return on the asset, and perform better for the client. For example, the Canadian architect, Frank Gehry, when he built the Guggenheim Museum in Bilbao, was able to create a totally innovative, landmark building because his design process was tightly linked through IT to his suppliers. This meant that he was able to ensure the feasibility of his design as he developed it.

4.2 IT Use

We found widespread agreement throughout the industry that IT is necessary and valuable. This industry view is based on the gains firms have made through automation, and on the perception that "ours is an information intensive industry." Typically, firms in each industry sector have adopted IT systems and tools which directly assist in the performance of their specialist tasks. This has allowed them to automate a number of time-consuming and error-prone activities and gain Level 1 benefits in cycle-time, productivity and accuracy. For example, the use of CAD for drafting has resulted in firms in all sectors gaining these benefits when changes to drawings are required. This usage reflects the automation phase of IT adoption. Its benefits have been achieved with little organizational change.

So far, the inter-organizational application of IT in the building and construction industry has delivered only Level 1 benefits. It has been confined to the automation of communications. Larger contractors typically use PCs to relay progress information from sites to the head office, and occasionally for some intra-firm knowledge sharing. Inter-firm e-mail partially replaces ordinary mail, courier, telephone and fax, and it is common for CAD files to be e-mailed to and from consultants. One architecture firm gave some consultants read-only privileges to access its CAD files, but this is rare. Electronic Funds Transfer (EFT) automates payments among contractors. These innovations have helped reduce cycle time and are now regarded as a competitive necessity.

Some firms we interviewed had begun to "informate" (Zuboff 1988). Further, in a few of the firms studied, the use of IT had begun to transform the way they do business. For example, the architectural firm of Flower and Samios chose to integrate the architectural design and documentation processes, thereby eliminating the need for draftsmen (Yetton, Johnston and Craig 1994). That firm also discovered that it can add greater value for its clients by providing them with a wider range of services based on the new core competencies it has developed. Companies like this have succeeded in staying ahead of their competitors not merely by automating but by changing their organization as well. Their strategic advantage has been their preparedness and ability to continually innovate, and to manage the change necessary to gain substantial business benefits.

We encountered a very few examples of initiatives toward achieving Level 2 benefits. One major principal contractor, Civil and Civic, had established a project-specific Web site on which all specialist contractors would record progress and other relevant information. The industry's largest client, the New South Wales Department of Public Works and Services (DPWS), had started to require that their contractors employ project-centered databases on contracts over a certain size.

4.3 Barriers to IT Use

By comparison with other industries such as retail and financial services, building and construction has been slow to adopt IT and slow to exploit it to its maximum. Reasons for non-adoption of IT include:
* Firms are very cautious in relation to technology risk and the risks associated with organizational change.

- Industry profit margins are generally so tight that firms, especially smaller ones, do not feel able to invest in change.

Reasons for slow or limited adoption include:

- The industry's emphasis on cost reduction means that it has little interest in benefits beyond automation.
- The benefits of automation for individual firms have been substantial, and for many may have seemed sufficient.
- Many firms believe that IT is a one-off investment.
- Fear of over-investment was widespread with several interviewees telling the researchers stories of companies actually or nearly going bankrupt as a result of spending too much on IT.

Unfortunately, reluctance to adopt new technology has resulted in some non-adopters, particularly architects, to go out of business. Those firms that have led in automation have gained business benefits such as greater volume of business and undiminished profit margins for a short period while their competitors have caught up. However, in the case of automation, it is so easy to imitate competitors' successes that over time costs have been reduced across the board and IT-based automation has become a competitive necessity but not a sustainable advantage.

In addition to those more general inhibitors of IT use, this study identified a number of inhibitors specific to IT-enabled collaboration. These included:

- Lack of vision. Few companies or senior executives have an overarching vision for the industry. Few are aware of the opportunities afforded by IT-enabled collaboration. The current Olympics-based boom encourages a view that the industry is doing well enough not to need to explore new, risky opportunities.
- Level of risk. Technology leaders saw themselves as potentially disadvantaged by having to bear costs associated with collaborating with less mature users of IT.
- Lack of a business model. Unlike auto manufacturing or retail-supplier relationships, there is no known example of how such a collaboration can be conducted satisfactorily in this industry. The industry does not have a basis for successful collaboration. Its culture is adversarial with a consequent lack of trust among firms. Given the struggle of industry players to enter business in other parts of the supply chain, it is not easy for potential collaborative partners across sectors to see how they might share the benefits of collaboration without risking their competitive advantage by exposing their operations to one or more partners. Many were concerned that collaboration would result in reduced margins as transparency would permit clients to more easily validate costings against actual work, and likewise contractors could better scrutinize subcontractors. Others were concerned that if the benefits do not accrue to competitors, they will ultimately all flow through to the customer. The risk-takers will not harvest the reward.
- Lack of capability. Automation through personal productivity technologies can be achieved without sophisticated IT management competencies. Inter-firm IT-enabled collaboration requires not only the ability to secure technology agreements with other firms, it also requires a level of managerial sophistication that is rare to find in this industry if knowledge sharing is to be achieved where previously there has been mistrust.
- Resistance to change. Resistance to the reengineering/organizational change necessary to gain a pay-off from the technology was commonly cited. Some

mentioned client resistance, clients being seen as highly conservative and uninterested in what IT might offer them. Technologists mentioned that senior partner and senior manager resistance was often based on lack of understanding of potential opportunities.

- Lack of a technology standard. While there are software technologies available to enable knowledge sharing and collaboration, none is currently regarded as standard.

In summary, up until now the industry has taken the easy gains from automation but these have typically not resulted in sustained advantage, merely survival. A few firms have gained more enduring competitive advantage through continual technological and organizational innovation. Even they have rarely innovated outside the boundaries of their own organization.   Consequently, the industry's biggest opportunity, the reengineering of the supply chain, is currently a largely unexplored challenge.

5. Discussion

The current competitive dynamic in the industry is that firms adopt IT as a necessity to drive down costs. Those who lag experience reduced margins or go out of business. The gap between the potential of the technology and its current use, even by industry leaders, indicates that these competitive dynamics will continue for the foreseeable future. However, unless the market grows significantly, the long run effect will be a decline in the number of firms in the industry and in the number of people employed, with the risk that (1) the industry will have limited opportunity to internationalize because current firm-specific cost reduction does not confer sufficient advantage for expansion into overseas markets and (2) overseas firms might successfully enter the Australian market.

The industry has the opportunity to mitigate some of these effects if it can be successful in learning how to reconfigure the supply chain and transform its organizations so as to deliver new, improved and qualitatively different service. In the remainder of this section, we discuss the two most likely organizational forms by which a new supply chain might be realised. We then make some recommendations for governmental or industry-level initiatives to help remove some of the barriers and kick-start the search for a reengineered supply chain.

5.1 Organizational Implications

Under the existing industry structure, the most obvious model for achieving a reengineered supply chain is a strategic alliance among several firms, each in a different sector, who would collaborate to develop integrated IT-enabled inter-organizational processes for their mutual benefit. The benefit would derive from their being able to better conduct business among themselves and by being able to compete more successfully for a larger range of projects.

The advantage of the strategic alliance model is that the benefits of specialization are fully retained. All parties retain economic incentives to innovate. There is flexibility to partner with firms other than those in the strategic alliance. The disadvantage is that the industry lacks a business model. This lack is particularly important because of the scarcity of trust in the industry, and the perceived levels of risk. In view of the power struggle among the consultants and contractors, it is even difficult to see just two players agreeing to share information and knowledge at sufficient depth to gain substantial benefits. Client preferences for varied mixes of consultants and contractors will militate against the creation of stable alliances. Moreover, because of the punctuated nature of industry relationships, there would be no guarantee that two collaborative partners could immediately build on an initial, experimental collaboration. The lack of a common technology platform or inter-operability standard will militate against experimenting with alliances.

A more likely scenario involves the creation of a vertically integrated "design and construct" business. The firm that vertically integrates will be able to explore different configurations of the supply chain because it has managerial control of multiple specialisms. It will, therefore, be able to explore and develop ways of integrating knowledge from different sectors. For example, it will be easier to ensure that construction knowledge is integrated into designs. While IT provides the communication technology to achieve such coordination, the firm will still need to evolve processes by which multiple sets of expertise are effectively pooled.

The advantage of the vertically integrated organization is that multiple industry specialisms are brought under the control of a single business vision. There is less reason for the specialists to compete and more reason to collaborate. It is flexible in its scope in that it is not necessary to attempt to integrate all parts of the supply chain. It is flexible to establish in that it is possible to integrate parts of the supply chain incrementally. Once established, it is likely to be stable and durable, permitting the organization to exploit the potential of cross-sectoral IT. The disadvantage of this form of organization is that it risks individual specialisms losing their focus, and hence their leadership in a particular area of expertise, particularly if the dominant form of management becomes process-based rather than functionally-based. This form of organization requires active management to oversee the introduction of new business processes and new management processes. It needs management to break down deeply ingrained prejudices and rivalries among staff from the different parts of the industry. It may encounter resistance from clients who are accustomed to being able to specify which architects, which consultants and which contractors are employed.

5.2 Recommendations for Government Supported Industry-level Action

A new industry supply chain will emerge from the commercial decisions of one or more firms. The task for government and industry bodies is to dismantle some of the barriers to IT use and to help enable technology-based collaboration.

The most basic requirement for action is the need to build awareness of the value of IT and to enhance and diffuse IT skills throughout the industry. Awareness can be built through a campaign to communicate the benefits of IT. For the purposes of preparing the ground for collaborative technology, it is important to emphasise this link between the competitive benefits of IT and organizational change.

This could be partly achieved by the government and industry sponsoring research to develop a body of case studies. In view of the hierarchy of subcontracting in the industry, in order to reengineer the supply chain it will be necessary to build the IT skill base of SMEs which currently make little or no use of the technology.

Therefore, government support for a program of short courses which provides both the business motivation for SMEs to invest in IT and the skills to use it is recommended. These courses could be run through a range of government sector technical and educational institutions with strong involvement from respected industry figures. A third initiative government and industry could take would be to develop and support a bank of information on industry best practice in relation to construction industry use of IT.

In terms of developing the opportunity for cross-sectoral collaboration, the industry needs to know about whatever business models have been adopted internationally and what impact they have had on industry structure and supply chain processes. Research leading to a published report would help build awareness of the opportunity across the industry as a whole.

Governments can also help develop the cross-sectoral opportunity through their role as major customers. By developing appropriate tender guidelines and contract management procedures, federal and state governments can specify varying degrees of collaboration as a contract requirement. The New South Wales Department of Public Works and Services (NSW DPWS) introduced such a requirement in 1998.

A third area for industry-level intervention is in developing a more value-added focus with attention to enhancing performance throughout the lifetime a building or facility. If a reengineered supply chain is going to deliver a sustainable advantage to firms, they will have to extend their thinking beyond cost and cycle-time reduction. The opportunity afforded by collaborative technology is for consultants and contractors to help add value for developers, occupants, operators and property managers by making buildings more usable, more appropriate to the needs of occupants, more flexible among uses, more manageable and more maintainable. While the top contractors have already started to adopt such a perspective, industry and government can accelerate its diffusion and acceptance by establishing industry awards for IT-based innovation and by supporting industry forums to explore collaborative value-adding strategies.

Finally, while collaborative technologies are already available and in use in isolated cases, government and industry should support efforts to establish an acceptable standard for collaborative IT in building and construction. Two initiatives stand out as worth supporting. First is the establishment of project-centered shared databases through which all participants in the supply chain can provide and access information about project progress. This is what NSW DPWS is promoting for projects where it is the client. Second is support for initiatives to achieve inter-operability among IT platforms. Industry technology leaders are currently involved in international standards bodies and they should be financially supported and encouraged to expedite their work.

6. Conclusions

The Australian building and construction industry has an opportunity to increase its competitiveness through reengineering the supply chain to increase information and knowledge sharing and collaboration. This has the potential to cut costs, reduce cycle time and add new value. However, the adoption of appropriate IT is a prerequisite and, as this study shows, a number of factors inhibit such adoption. This finding is made all the more pertinent by the fact that this particular industry has a long history of bringing a number of firms together into a virtual organization for the duration of a project. So, despite concerns about the potential threat of an international firm devising a business model for collaboration down the supply chain and entering the Australian market, and despite awareness that for the Australian industry to be the first mover would give it the opportunity to achieve a powerful position in overseas markets, it is doubtful that market forces alone will result in rapid redesign of the supply chain.

In order to reduce the risk of the industry being overtaken by foreign competition, initiatives are required at both the industry and governmental levels, although these can only help remove barriers: they cannot mandate new inter-organizational processes except insofar as government in its role as a powerful client can impose specific requirements in its contracts. Even then, while such requirements may impose some information sharing and require the adoption of some common technology, strategic alliances will not be established and significant knowledge will not be shared until individual consultants and contractors have a business model that offers mutual benefit and some protection from risk.

Creating the conditions for an IT-enabled supply chain is an appropriate part of government's general role in industry development. Nevertheless, it creates a dilemma. A reengineered supply chain will involve a degree of industry restructuring including rationalisation and probably consolidation. Some firms will be put out of business. So government will be faced with bearing the political cost of business failures in the interests of maintaining the industry's global competitiveness. In an industry consisting of so many small and vulnerable firms, the political fallout could be serious, so government may need to consider how to soften the landing for the losers in the industry restructure.

More generally, this study has shown that IT-enabled knowledge sharing will not be readily achieved by an industry just because inter-organizational technology is available and there are potential business benefits. The barriers can be substantial and may need to be addressed at the industry level. It is therefore necessary to understand more than the balance of benefits and costs for individual firms; it is necessary to understand how existing industry factors affect the likelihood of adoption. In particular, in an industry structure where no group of players is so powerful that it can impose knowledge sharing and collect the bulk of the benefits itself, market forces alone are unlikely to be sufficiently compelling for industry players to choose collaboration.

References

Ahmad, I.; Russell, J.; and Abou-Zeid, A. "Information Technology (IT) and Integration in the Construction Industry," *Construction Management and Economics* (13:2), 1995, pp. 163-171.
Bensaou, M. "Inter-organizational Cooperation: The Role of IT. An Empirical Comparison of U.S. and Japanese Supplier Relations," *Information Systems Research* (8:2), 1997, pp. 107-124.
Brandon, P.; Betts, M.; and Wamelink, H. "Information Technology Support to Construction Design and Production," *Computers in Industry* (35:1), 1998, pp. 1-12.
Chatfield, A., and Yetton, P. "Moderating Effects of EDI Embeddedness on Time-based Strategic Capabilities," in *Proceedings of the Sixth European Conference on Information Systems*, W. R. J. Baets (ed.), Aix-en-Provence, France, 1998, pp. 853-869.

Choudhury, V. "Strategic Choices in the Development of IOIS," *Information Systems Research* (8:1), 1997, pp. 1-24.

Damsgaard, J. "Global Logistics System Co. Ltd.," *Journal of Information Technology*, Forthcoming, September 1999.

Davidow, W., and Malone, M. *The Virtual Corporation*, New York: Harper Business, 1992.

Hart, P., and Saunders, C. "Power and Trust: Critical Factors in the Adoption and Use of EDI," *Organization Science* (8:1), 1997, 23-42.

Janah, M. "Wal-Mart Links to Suppliers," *Informationweek*, October 5, 1998.

Johnston, H., and Vitale, M "Creating Competitive Advantage with Inter-organizational Information Systems," *MIS Quarterly*, June 1988, pp. 153-165.

Konsynski, B., and McFarlan, W. "Information Partnerships: Shared Data, Shared Scale," *Harvard Business Review*, September-October 1990, pp. 114-120.

Kumar, K., and van Dissel, H. "Sustainable Collaboration: Managing Conflict and Cooperation in Inter-organizational Systems," *MIS Quarterly* (20:3), 1996, pp. 279-300.

Loebbecke, C.; van Fenema, P.; and Powell, P. "Knowledge Transfer Under Coopetition," in *Information Systems: Current Issues and Future Changes*, T. Larsen, L. Devine and J. DeGross (eds.). Laxenburg, Austria: IFIP, 1999, pp. 215-229.

Quinn, J.; Doorley, T.; and Paquette, P. "Technology in Services: Re-thinking Strategic Focus," *Sloan Management Review*, Winter 1990, pp. 79-88.

Snow, C.; Miles, R.; and Coleman, H. "Managing 21st Century Network Organizations," *Organization Dynamics* (20:3), Winter 1992, pp. 5-20.

Stalk, G.; Evans, P.; and Schulman, L. "Competing on Capabilities: The New Rules of Corporate Strategy," *Harvard Business Review*, March-April 1992, pp. 57-69.

Yetton, P.; Johnston, K.; and Craig, J. "Computer-aided Architects: A Case Study of IT and Strategic Change," *Sloan Management Review*, Summer 1994, pp. 57-67.

Zuboff, S. *In the Age of the Smart Machine*. New York: Basic Books, 1988.

About the Authors

Chris Sauer is Senior Research Fellow in the Fujitsu Centre at the Australian Graduate School of Management. He researches IT project risk and project management, and issues relating to IT-based organizational transformation. His latest book, co-edited with Philip Yetton, *Steps to the Future: Fresh Thinking on the Dynamics of IT Based Organizational Transformation*, was published by Jossey-Bass in 1997. He is Deputy-Chair of IFIP WG8.6, Asia-Pacific Editor of the *Journal of Strategic Information Systems* and a member of the editorial board of the *Information Systems Journal* and the *Australian Journal of Management*. Chris is the contact author for this paper and can be reached by e-mail at C.Sauer@unsw.edu.au.

Kim Johnston is Research Fellow in the Fujitsu Centre at the Australian Graduate School of Management. His research interests are in information technology and organizational design. He has published a number of articles in books and journals such as *Sloan Management Review*. Prior to joining the Fujitsu Centre, he worked as a management consultant in Sydney.

Khalid Karim is Research Engineer in the Australian Centre for Construction Innovation in the Faculty of the Built Environment at the University of New South Wales. Previously he has worked as an engineer on a number of large scale industrial and infrastructure projects. His recent publications have focused on benchmarking in the Australian construction industry.

Marton Marosszeky is director of the Australian Centre for Construction Innovation in the Faculty of the Built Environment at the University of New South Wales. Current projects include performance measurement and benchmarking; waste management in the housing sector; risk management in occupational health and safety; and the study of quality related issues in the residential construction sector. Before founding the Building Research Centre in 1987, he worked as a structural designer of tall buildings. He has published numerous academic and industry research papers and provides a range of advisory services to industry.

Philip Yetton is the Commonwealth Bank Professor of Management and the Executive Director of the Fujitsu Centre for Managing Information Technology in Organizations at the Australian Graduate School of Management. His major research interests are in information technology, leadership style, and decision making. He is co-author with Professor Victor Vroom of the internationally acclaimed text *Leadership and Decision Making*, and is co-author of the leading text book *Management in Australia*. His latest book, co-edited with Chris Sauer, *Steps to the Future: Fresh Thinking on the Dynamics of IT Based Organizational Transformation*, was published by Jossey-Bass in 1997. He has written numerous research papers and is on the editorial board of *Leadership Quarterly* and *Organizational Science*.

11 ORGANIZATIONAL DISPOSITION AND ITS INFLUENCE ON THE ADOPTION AND DIFFUSION OF INFORMATION SYSTEMS

Brian O'Donovan
University of the Witwatersrand
South Africa

Abstract

This study proposed the concept of organizational disposition as a means to increase the understanding of the dialectic nature of the adoption and diffusion process. The concept of organizational disposition was incorporated into an integrative framework, which provided the theoretical background for two case studies. In both studies, the reactions of the users brought about a dialectic process in which the moods experienced enabled the opening up or uncovering of the reasons for resistance. The restoration of the common sense of the organization was reached through an ongoing process which integrated and reconciled the users' and implementers' dispositions into a coherent whole.

Keywords: Heidegger, disposition, adoption, diffusion, implementation, user reactions.

1. Introduction

Adoption can be described as the initiation, design, development and installation of an information system, and diffusion as the subsequent spread of the system through the organization and its incorporation into the organization's routines (Kwon and Zmud 1987; Prescott and Conger 1995). The terms adoption and diffusion suggest two distinct stages but this is essentially one overlapping and interacting process. This process can result in organizational upheaval accompanied by an unenthusiastic reception and resistance, resulting in potentially useful systems being rejected or abandoned (Wastell and Newman 1996).

There is insufficient understanding of the reasons for the frequent failure of managers to initially adopt or subsequently to diffuse information systems into organizations (Markus and Benjamin 1996). The search for these reasons can be enhanced by a better understanding of the nature of organizational change brought about by new information systems and the reactions to that change (Wastell and Newman 1996). This study seeks to increase this understanding by introducing the concept of organizational disposition to explain how reactions to the adoption and diffusion of an information system come about. The concept is also used to explore how the disruption of a proposed information system is resolved and how the organization is restored to a cohesive whole with or without the successful diffusion of the system.

The study will proceed as follows. First, a view of the nature of organizations and of change in organizations is put forward. This view is influenced by Heidegger and introduces the concept of organizational disposition. Using this concept, the strategies of implementers, the reactions of users, and the process to restore a common sense of the organization are brought together in a framework. Finally, the framework is used to explain some of the dynamics of this process using two case studies.

2. The Nature of Organizations and of Change in Organizations

Heidegger's exposition of *Dasein* addresses the primordial or essential being of the human person. The term *Dasein* is used by Heidegger to describe existing (*Sein*—being) within (*Da*—both here and there) a world where others exist (Introna 1997, p. 27). Or more simply put, the human being as a "being" that is always already "there" (present), that is always already involved and engaged, as its constitutive essence. In Heidegger's analysis *Dasein* does not describe the nature of communal groups such as organizations. However, it is contended that organizations share in the nature of *Dasein* but in a way that can be described as a subpattern within the grand pattern that is *Dasein* (Haugeland 1982, p. 19).

This subpattern draws on the context in which the organization is embedded and in which there operates a generally accepted set of norms and conventions that become part of what it means to do something in that world called the organization. Heidegger uses the term *Das Man* (which can be translated as the anyone) to describe this set of shared norms and conventions. It refers to the impersonal and non-specific "they" in phrases such as "they said it" or "they don't like it." Each organization, in turn, establishes its own local *Das Man,* which enables its ongoing activity to take place with a reasonable amount of order.

This local *Das Man* can be referred to as the cultural structure (cf. Walsham 1993, p. 34). It incorporates culture in the customs and practices that develop over time and provide a web of collectively accepted meanings and assumptions, which becomes the frame of reference underlying the perceptions, interpretations, understandings and communications of the members (cf. Schein 1990). It also incorporates structure in the way the members undertake their day to day activities and interact accountably, guided and controlled by common norms, conventions and rules, and the currently accepted distributions of power.

The cultural structure underlies all organizational activities and binds and articulates those activities. These activities, in turn, require the use of objects such as tools and materials, or as Heidegger terms it, "equipment." In using "equipment," organizational members encounter others who are relevant to their use of that equipment. The interconnected network of "equipment" is thus placed within a context of activities in which others are involved. It is those others who add human purposiveness and give the organization meaning or significance. The organization is thus made up of a cohesive whole of a context of meaning or significance bound together by the cultural structure.

There are three influences on the maintenance of the cohesiveness of the organization: socialization, social networks, and conditions of work. Socialization, begins with procedures to attract and select individuals whose attitudes or values are likely to be congruent with organizational needs and continues with formal and informal training and direction (Van Maanen and Schein 1979). New members will be susceptible to clues and signs given to them by those who support and guide them in their roles and demonstrate the attitudes and values specific to the organization. These actions include ostracizing those who do not fit in or accepting those who do. Through social networks, the attitudes and values of the members are further influenced by their perceptions of the nature of other groups and organizations and the social and historical context in which the organization is embedded.

The influence of conditions of work reflects the need for a reciprocal balance between the members of the organization. This reciprocal balance is not a conscious agreement but arises from the way individuals cooperate with others to achieve their own goals and expectations. Members take on obligations in exchange for tangible benefits, such as material reward, and intangible benefits such as opportunities for advancement or a comfortable work situation (Silverman 1970, p. 178). Some members will confer power on others in order to have less responsibility and reduce uncertainty (Coombs, Knights and Willmott 1992). Others will seek responsibility in order to have power and prestige. Reciprocity is thus sustained by the acceptance of the distribution of power as well as the reciprocal commitments and entitlements that arise from membership of the organization (Gouldner 1960).

The recognition, understanding and acceptance of opportunities for mutual benefit provide the basis for the transformation of the organization. In his exposition, Heidegger puts forward four "ways of being" that presuppose and produce the shared nature of *Dasein*. These are *Dasein's* way of disclosing the world and discovering things in it (Dreyfus 1991, p. 166). If the four "ways of being" are adapted to the context of an organization, they describe how innovative ways to do things are uncovered, evaluated, find acceptance and become incorporated into the routine. The full discussions of these are beyond the scope of this paper and only the aspect of organizational disposition will be considered.

3. Change and Organizational Disposition

Change can be said to result from an interdependent progression of actions, reactions and interactions (Pettigrew 1987). A key aspect of this progression is the reactions of organizational members to change. When faced with possible changes to ongoing activities, the members have dispositions that influence whether they "accept" some or "reject" others ("accept" or "reject" does not refer to cognitive processes, as will become clear). Heidegger's "way of being" relevant to these dispositions is *Befindlichkeit*. *Befindlichkeit* has been translated as "affectedness" in the sense of a shared sensitivity to those people and things that matter. It has also been translated as "so-foundedness" in the sense of being "found" in a situation where certain possibilities are of importance (Dreyfus 1991, p. 168). It may be best described as that which we find ourselves *in*, when we suddenly stop to take notice (be it in a particular situation, condition, predicament, opportunity, and so forth). The important point is that we *always* find ourselves *in* something, entangled in a host of situations, predicaments, opportunities, and so forth—this is our "so-foundedness," our *Befindlichkeit*. In our finding ourselves *in*, some things matter and others do not, some ways of doing make sense and others do not. We are never merely neutrally "there." The way in which our *Befindlichkeit* manifests itself in our everyday dealings in the world is as mood (*Stimmung*). Here mood is not purely an emotive, internal, notion (such as "he is in a bad mood") but rather a particular sense or tacit awareness of our being *in*-ness, of finding ourselves in the world. It is that "just below the surface" sense of awareness of the importance (or not) of a particular moment (of a particular situation, condition, predicament, opportunity, and so forth) that we find ourselves *in*. As such, it discloses the world and its possibilities to us.

In the organizational context, *Befindlichkeit* will be referred to as organizational disposition. Organizational disposition is not a state of mind (despite the use of the term by the translators of Being and Time), nor is it a cultural disposition to behave (Dreyfus 1991, p. 168). It is not the additive combination of individual dispositions, but the common moods of the members which are reflected from the organizational world rather than from introspection (Dreyfus 1991, p. 174). It is a tacit awareness, which flows from being engaged in the world, of a collective "finding ourselves in" that is just there but never really *there*—a sort of collective sense of "being in this thing together" and a tacit awareness of what it means in terms of expectations and actions. "It comes neither from "outside" nor from "inside" but arises out of being-in-the-world, as a way of such being" (Heidegger 1962, p. (176)[136]).[1]

As a way of being, organizational disposition is a way of disclosing or uncovering the nature of a current situation in the cohesive whole of the organization—both the context of significant or meaningful activities and the cultural structure. Through disposition, the members have an attunement or sensibility to everyday situations. This attunement is manifested in moods which have already disclosed being-in-the-organizational-world as a whole (Heidegger 1962, p. (176)[137]). One could say that the organizational disposition is the *common sense* of the members in the situation. Common sense should be understood in two ways. First, common sense refers to the fact that it is a mood, a sense, of what it, the whole (the situation), is "telling" us. Second, it

---

[1] The practice of using two page numbers to reference quotations from *Being and Time* will be followed. The first page number in parentheses refers to the standard English translation and the second page number in brackets refers to the page in the standard German. It should be noted that many of the cited references to Heidegger and commentators on his work refer to the individual *Dasein* and these have been adapted to apply to organizations.

refers to the fact that it is an everyday tacit awareness of what would be the appropriate response in *this* particular situation in which we find ourselves.

The members are always already surrounded by the cohesive whole of the organization and this whole matters to them as, within it, they have a capacity or ability to be. There is thus always the possibility that members will find opportunities or new ways to do things. There is also always the possibility that members will find themselves in certain moods as a result of confronting these opportunities or new ways to do things. These moods reveal how things are going for the members and whether these things matter (Dreyfus 1991, p. 172). Through these moods, organization disposition attunes the members' tacit awareness of the nature of change and "informs" them of appropriate responses to this change (Heidegger 1962, p. (172)[134]). Organization disposition is thus a way of being which affects the members' understanding of the nature of changes to activities. It is at once a sensibility, a sensitivity and a sensing—a prevailing common sense.

It is important, and hopefully clear from the above, that organizational disposition is not the culture of the organization but rather the situated attunement (sense of the whole) that flows from engaged action in the world. One may argue that it is an element of culture. However, the discourse on culture (even in Schein) has tended to turn culture into a "thing" (in the minds of the organizational actors)—to be measured, managed, and so forth. This sort of conception perpetuates the Cartesian dualism of separating cognition and action. In this sense, one could say that organizational disposition is a post-Cartesian notion of organizational culture. If one wants to retain the notion of culture (as a post-Cartesian notion) then it would best be describes at the background or horizon that renders certain dispositions possible and others not. The claim of this paper it that it is disposition (as attunement) that is of real importance in the implementation of information systems. As such, it has a substantial impact on the process of the adoption and diffusion of an information system. The discussion of this impact will first consider the perspectives of implementers and users before discussing how these actions subsequently lead to a restoration of a common feeling of sense.

4. Implementers and Users

Most organizational changes rely on the relative ability of some individuals or groups to impose their expectations on others (Berger and Luckmann 1966, p. 101). However, in the dialectic of control, most individuals or groups have a degree of power to influence the nature of change (Giddens 1984, p. 16). Any individual or group may present possibilities for change provided that they can form alliances and obtain control over the resources needed to achieve the required outcomes. Those without the power to propose changes still can accept or resist them (Frost and Egri 1991).

The process of change brought about by the adoption and diffusion of an information system can thus be viewed from two perspectives: that of "implement-ers"—not just the sub-group of designers, developers, and installers, but also those that support the system; and that of "users"—the subgroups that will use and be affected by the new system. The next section will commence the discussion of these perspectives with the strategies of implementers to present opportunities or new ways of doing things.

4.1 The Strategies of Implementers

Implementers attempt to reveal things as doable, as making sense in the cohesive whole of the organization (Dreyfus 1991, p. 185). In this attempt, they use interpretations and assertions. Interpretations involve the uncovering of the cohesive world of the organization in order to appropriate the change. Assertions point out, elaborate and communicate the nature of the change (Dreyfus 1991, pp. 208-211). Often, implementers focus on the assertion of the value of the change such as the impact of information technology on the nature of purposeful activities. Less frequently, the focus will be on the interpretation of its impact on the underlying cultural structure. In so doing, implementers will draw on socialization methods and approaches, seek to change the reciprocities and balances of power in the ongoing conditions of work, and incorporate those methods and technologies which are approved or fashionable within their social networks.

Implementation strategies go beyond the specification, design or selection of the system and include the formation of alliances, obtaining control over the required resources, and preparing people, facilities, structures, and processes to accommodate the change. These strategies are elaborated in three categories derived from the work of DiMaggio and Powell (1983) and Fogarty (1992).

Organizations are modified by forces in their organizational field which result in them becoming more similar to one another, a process DiMaggio and Powell term isomorphism. They identify three mechanisms through which institutional isomorphic changes occur: coercive, mimetic and normative. Fogarty suggests that internal organizational socialization processes can be classified into three similar interacting and overlapping influences. These three isomorphic influences have, in turn, been adapted to arrive at three categories of implementation strategies.

4.1.1   Coercive implementation strategies

Coercive strategies are authoritarian, adversarial and imposed (Wastell and Newman 1996). They rely on the recognized authority and power of the implementers and the citing of rules and procedures. When the users have little opportunity to leave the organization, for example, when there is high unemployment, coercive strategies are more likely to be used. If the users have the possibility of mobility, and the proposed changes compare unfavorably with other organizations, implementers can place less reliance on a coercive approach.

Coercive strategies include technical training and indoctrination, direct supervision of the usage of the system, and reward for acceptable behavior. They can also use hidden inducements such as forcing those who refuse to use the system to produce the required information manually (Markus 1983).

4.1.2   Mimetic implementation strategies

Mimetic strategies occur when there is less clarity over who has the power to implement and when there is a necessity to build alliances in order to proceed. If implementers can persuade enough members to accept the system, actions and perspectives that are considered appropriate and desirable by the majority will be promoted. These strategies encourage members to adopt the change through the use of workshops, demonstrations and "ideological training," which emphasizes features valued by the users such as modernity and ease of use (Newman and Sabherwal 1996). These strategies are influenced by the perceptions the users have of similar systems in other organizations. Other techniques include mentoring and role modeling with the intention of getting the members to identify with and commit themselves to the new system.

4.1.3 Normative implementation strategies

As the users in a normative environment usually have a high level of power, it is likely that they will require considerable involvement in the entire adoption and diffusion process (El Sawy 1985). Normative strategies are designed to show that the system will contribute to the success of both users and the organization. Normative strategies include demonstrating that the system does not conflict with existing cultural values, standards and norms (El Sawy 1985). Other strategies include appeals to users to demonstrate their sense of loyalty and commitment (Frost and Egri 1991). When normative strategies are supported by the acceptance of respected peers, this can encourage acceptance of the system. However, senior managerial support is less important in this environment (Prescott and Conger 1995).

4.1.4 The combination of implementation strategies

The combination of the strategies used will be influenced by the implementers' levels of power or support and their degree of confidence, experience and skills (Frost and Egri 1991). The greater these are, the more likely that coercive strategies will be used. Strategies will also be influenced by the degree of change introduced by the system. The greater the change, the more resistance can be anticipated and the less likely coercive approaches will be used. In addition, each successive stage of the adoption and diffusion process is conditioned and influenced by previous events and actions (Walsham 1993, p. 57). Implementers will often be alert to this ongoing process and adapt their strategies accordingly.

Implementation strategies are not intended to overcome resistance but avoid it, if possible, and confront it constructively if not (Markus 1983). A key objective is to gain the support of users by fostering enthusiasm and commitment for the change rather than compliance (Senge 1990, p. 9). This requires an understanding of the possible reactions of users, the subject of the next section.

4.2 The Reaction of Users

The adoption and diffusion of an information system causes a form of breakdown—an upset to established ways of doing things. This breakdown brings to the fore the users' innate and historical dispositions, which are manifested as moods. As the members care whether the proposed change affects them and the achievement of their goals, they search for the reasons for the moods (Dreyfus 1991, pp. 175). Organizational disposition as a way of being opens up or discloses the impact of the change on the cohesive whole of the organization. The outcome of this is the reactions of users.

When the users can marshal sufficient power to resist a proposed system, their reactions remain primarily on the surface, involving detailed disclosure and explanation. The less the power of the users, the more likely that the strategies will be deep strategies, intense, political and sometimes irrational (Frost and Egri 1991). Reactions can thus vary in nature and intensity. These reactions will be discussed under three headings: rejection and failure, resistance and modification, and acceptance and commitment.

4.2.1   Rejection and failure

Reactions to a new information system can include treating the current ways of doing things as inviolate and thus ignoring or refusing to use the new system (Frost and Egri 1991). An alternative rejection approach is to ensure that the system does not function by deliberately using it incorrectly or ensuring that the output is incorrect and unusable (Myers 1994; Wastell and Newman 1996). Users may thus contend that they cannot use a system for technical reasons rather than disclose the true reasons for resistance (Markus and Keil 1994).

Users may be unhappy with the system but fear to directly resist the change. They may also not have sufficient power to reject the system. In this case, the reactions will attempt to modify the process.

4.2.2   Resistance and modification

Resistance and modification reactions are calculated to prevent the successful use of a portion of the system, or to customize the system to reflect the users' own requirements. Tactics include modifying the system to operate as if no change had occurred and keeping alternative sets of records (Markus 1983; Tyre and Orlikowski 1994).

Even when the users ostensibly accept the change, they may use subversive tactics to resist and modify. This can particularly apply when the implementation method has been coercive and the users feel anger at the way the change is implemented (Wastell and Newman 1996). Any resistance under these conditions will be likely to use informal power. Attempts will be made to build coalitions and gain support of other interest groups in order to collectively resist change (Frost and Egri 1991). At times, users will lack the collective organization to resist or they may be constrained by fear of reprisals (Clegg 1989, p. 220).

4.2.3   Acceptance and commitment

If the new system does not result in much change to activities and, in turn, to the norms, rules and meanings already institutionalized in the organization, there is a likelihood that users will accept and commit themselves to the system (Introna 1997, p. 134).

Once the attempts to reject or resist and modify the system have been resolved or if there is substantial acceptance and commitment, the organization consolidates into a cohesive whole again. The next section will discuss the restoration of the common sense of the organization through a dialectic interaction between the interpretations and assertions of the implementers and the dispositions of the users manifested in reactions.

4.3 Restoring the Common Sense of the Organization

The diffusion of the information system into the ongoing everyday work of the organization will only be complete when the common sense of the organization is restored. This state is reached through an ongoing process which integrates and reconciles the users' and implementers' perspectives into a coherent whole. It may be that this coherent whole is quite unlike that anticipated by either the implementers or the users. Nevertheless, this, for the time being, will enable the organization to settle back into everyday activity. The process of restoring this common sense of the organization is illustrated in the framework developed below.

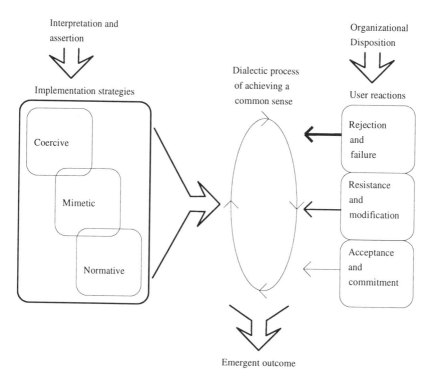

Figure 1. A Framework to Illustrate the Dialectic Process of Restoring the Common Sense of the Organization

The historically based dispositions toward change trigger the evaluation of the possibilities presented by the implementers. Over time, users will attempt to manifest the reactions of rejection, resistance and modification, or acceptance and commitment, shown in Figure 1 as three possible user reactions which can influence the process of restoring the common sense of the organization.

Implementers' interpretations and assertions result in attempts to persuade and influence the users to accept the change. This is directed toward members re-evaluating their attitudes toward change and a reduction of resistance and confusion. If the implementation strategies fail, the system may have to be modified. The use of combinations of the implementation strategies are illustrated by the three overlapping strategies in the block on the left of Figure 1 and the arrow combining those strategies to influence the process of restoring the common sense of the organization.

There is a creative tension between what the implementers propose and the users' perception of what they want. This tension has been described as the dynamic equilibrium between holding on and letting go (Kofman and Senge 1995) or between persistence and change (Robey and Azevedo 1994). As Heidegger puts it: "One only is by directing oneself to a future on the basis of an existing past" (1962, p. (41)[20]). The resolution of this tension is illustrated by the circular dialectic process in the center of the framework.

As the adoption and diffusion process continues, efforts to restore the common sense of the organization need to adapt to ongoing events in the organization. There are often modifications to the system to overcome technical problems or to better achieve the outcome the system is required to produce. The users may experience a general distrust or lack of motivation engendered by other events in the organization. These factors could introduce a new set of reactions from the users. Consequently, the possible combinations of implementer strategies and user reactions are complex and the outcomes are not easy to predict even with the same information technology (Sahay and Robey 1996). It is thus not surprising that the change brought about by the introduction of an information system is not only emergent but seems to be messy and unpredictable (Coombs, Knights and Willmott 1992; Zuboff 1988, p. 395).

The process of adoption and diffusion of an information system is thus complex and recursive. Because of this, the research into the application of the framework needed to be interpretive. In this study, case studies were used as these provided rich insights into the dialectic process of restoring the common sense of the organization and adopting and diffusing information systems.

5. Case Studies

The following two case studies are intended to explain how the interaction of the implementers strategies and users reactions as illustrated in the framework result in an emergent outcome. These can thus be categorized as explanatory case studies (Yin 1994, p. 3). Of the two companies, one had rejected a proposed system while the other had achieved success with the adoption and diffusion of a large system. The two companies thus gave insights into different aspects of the adoption and diffusion process.

The case studies used a variety of sources of data in order to put together a rich text for interpretation. Personal observation over a period of more than two years, while consulting to the respective organisations, established the overall context. The principal source of data was interviews conducted as part of a larger project. Thirteen interviews were selected as being of particular relevance to this study. However, the remaining 23 interviews provided a rich background to those selected. The interviews were conducted over a period of eight months and the average duration was just over an hour. They were tape-recorded and verbatim transcripts produced.

The framework developed above guided the semi-structured research questions asked. However, although the purpose was to confirm and expand the theory, it was considered important to be open to alternative theoretical explanations of the answers received. Thus each interviewee was assured of confidentiality and encouraged to express his or her own views and recollections about the adoption and diffusion processes. The interviews were supported by documentation such as annual reports, strategic plans, implementation reports, slide presentations and training material so as to strengthen the development of the theory by the triangulation of evidence.

The resultant data was examined from two perspectives. First, the nature of the implementer strategies and user reactions was established. These strategies and reactions represented the tangible, first level understandings of the adoption and diffusion process. Underlying these first level understandings were the organizational dispositions of the respective organizations. This required a further analysis of the data to identify occurrences of common moods, how these opened up the cohesive whole of the organisation and found what mattered to the users, and finally how this contributed to restoring the common sense of the organization.

The final step in the analysis of the case studies is to seek patterns across both cases and to compare those patterns with existing or "enfolding" literature so as to improve the possibility of generalisation (Eisenhardt 1989).

5.1 Bevco

Bevco is a South African company that manufactures a single type of product under different brand names. It is divided into four geographic regions which are further divided into the functional areas of production, sales and distribution, and marketing. The system investigated in this study, referred to as Saturn, was considered, almost accepted and then rejected in favour of an alternative system.

5.1.1 Implementation strategies

The proposal to purchase Saturn was driven by a single manager. He adopted a predominantly mimetic strategy, which attempted to market his proposal on the basis of satisfying regional information requirements. His strategy included frequent consultations with colleagues and presentations to management of the benefits of the system. In the initial stages, he appeared to have support from a small group of head office managers. However, they tended to treat his proposal as a catalyst for general discussion about the information system needs of head office rather than those of the regions. Despite considerable effort for nearly a year, a final decision was not taken. His frustration at his inability to get the head office managers to commit themselves was reflected in his description of them as

> *herds of the lateral thinking, intellectual guys who want to participate*
> *in these energetic, frenzied debates but there are very few people who*
> *are disciplined enough to go back and say: "What were we debating,*
> *what did we learn, what do we actually need to do as the next step.*

As it transpired, his approach failed, his support evaporated and he left the company.

5.1.2   User reactions

The main reactions to Saturn came from the finance group as it was having considerable difficulty in using its 27 incompatible legacy systems. When the financial director took over the information systems function, finance and information systems formed an alliance which set about putting forward two arguments to support their desire for better systems.

The first argument was that many of the financial systems were not year 2000 compliant. It was suggested that it was not practical to update the existing systems or to replace them with new systems compatible with Saturn. The proposal was to reject Saturn in favor of a system that would cater for the requirements of finance.

The second argument was the contention that the provision of financial information should be centralized in order to offer a better service. This was supported by a strategic need to buy raw materials centrally, a situation which had been signaled in the annual report. This need reinforced finance's argument as Saturn was unable to accommodate centralized procurement and finance systems.

An IS manager closely involved with the decision confirmed that, on the surface, these two reasons were accepted:

> *[W]e want to process our finance transactions centrally....[and] we*
> *want centralized procurement information....Those two things on their*
> *own effectively canned our previous decision to buy [Saturn].*

5.1.3   Restoring the common sense of the organization

The initial reactions to Saturn could be described as vague feelings that the proposal was not right. Initially, most of the managers interviewed reported a degree of discomfort with Saturn. These vague feelings and discomforts were to prove an important background to the process of restoring the common sense of the organization. It is informative to explore these further.

The manager proposing Saturn had a production background and had decided that the need was: *"[To give] tools to shop floor processes to help them solve problems and to make decisions."* This seemed to be a reasonable proposal as, over four years, a considerable degree of autonomy had been deliberately and carefully developed in the regions. This was intended to provide an environment where individuals and groups at all levels would be aware of and involved with Bevco strategies, be encouraged to translate those strategies into goals and to measure their progress toward the achievement of those goals. Saturn was to support this program by being installed at a regional level and being modified to suit local conditions. However, many head office managers felt that this autonomy had got out of hand and that a higher degree of centralisation was needed. This feeling was strongly expressed by the financial director:

> [M]y personal view is that people that tell shopfloor workers that they should be designing strategy are fools....Strategy should be designed by people who can see further than a year ahead.

This strongly worded statement pointed to a growing feeling, flowing from top management, that management information should be centrally provided. The managers were not always clear why this was needed and could only point to the need to buy raw materials centrally. This vague feeling was expressed by a manager of IS:

> Let me say one thing....we must run this thing centrally because we want central information...we want to keep this stuff centrally because it is a core business requirement, we have got to do that.

This need had been influenced by a management consultancy which had suggested, among other things, that the company as a whole becomes the unit of analysis rather than the regions. The idea was to allocate the company's resources where they would be the most effective and to introduce a greater level of central planning.

Underlying the mood of resentment toward the freedom given to the regions and the mood of discomfort with the lack of centralized information was the aggressive nature of interactions in the head office. A head office manager described how this aspect of the culture became apparent as a result of the managers' search for reasons for their reluctance to accept Saturn:

> [Bevco's] culture...may not be as chummy as it was in previous years. It is far more focused, far harder, far more business directed....It is highly individualistic and an internally very competitive culture.

As the implementing manager discovered, this had resulted in his proposal being subjected to intense scrutiny and discussion. This not only delayed the approval of the proposal and enabled resistance to develop, it also became acceptable to look for problems with Saturn.

A particularly strong feeling of resentment was expressed by the finance managers. They were upset that resources were being requested for a regional system when the existing financial systems were cumbersome and time consuming to operate. This mood provoked the finance managers into the realization that their conditions of work were unacceptable. This was a driver in their marshaling of acceptable reasons to justify their "resist and modify" reactions. This resulted in the finance group adopting the role of implementers and moving to the use of assertions. As the finance group was able to gain sufficient power to promote its needs ahead of those of the regions, they were able to reject Saturn in favor of an alternative system.

The common sense that eventually emerged was a rejection of the system destined for the regions and the acceptance of an alternative system that was essentially financial. The regions were not very concerned about this as they had developed a cynicism as a result of the many proposals from head office that had failed to materialize. Saturn was merely another example. Head office managers saw the decision as further enhancement of their power as the outcome would be more centralized information and finance had gained an improvement in their conditions of work.

5.2 Bankco

Bankco specialized in banking for individuals and small businesses and had more than 200 outlets throughout South Africa. The system investigated in this study processed the transactions generated in these outlets, facilitated the interface between the customer and the bank, and collected the data needed to update the branch accounting system on the mainframe. It was an internally developed system, referred to as the branch delivery system or BDS.

5.2.1   Implementation strategies

The decision to adopt the BDS was confined to a small group of head office managers. At the early part of any project in Bankco, it was unusual to consult with users. This coercive approach was accepted and provoked little reaction. The initial installation of the BDS was into a pilot group of eight branches. Due to technical problems, strong reactions began to develop soon after the pilot implementation. As a result, an independent organization was commissioned to conduct a post pilot implementation study (PPIS) of the impact of the system. This was followed up by an internal review. The PPIS and the review confirmed that there was strong resistance to the system and that there were design and other problems.

The outcome was the withdrawal of part of the system and a change in the implementation approach. The implementers learned to listen to users, to acknowledge the design problems, and to change the system where needed. Slowly, the implementation approach changed to introduce a mimetic element into a still predominantly coercive strategy.

After some time, there was a recognition by the users that the system had been accepted by a major proportion of the staff. When this point of critical mass was reached, the implementation strategies became more coercive. While staff continued to be included in the upgrading and rewriting process, this was felt to be tokenism rather than true participation.

5.2.2   User reactions

Reactions will be discussed in three stages. The reactions to the pilot implementation; the reactions while implementation was ongoing; and the reactions after critical mass was achieved.

The initial strong reactions. Technical problems, particularly poor response times, excessive down time, and an interface which made the system difficult to use, provoked requests to remove the new system and to revert to the previous one. There were particular objections to the impact of the system on the tellers. The staff said little at this stage. A branch manager in one of the pilot branches reported: "They were told: 'You will have the system and that's it'." However, the staff offered the PPIS a long list of minor faults. As they were unable to reject the system or even refuse to use the system, the only permissible official reaction was to find problems with the system.

As implementation proceeded, the initial strong resistance to the system waned. The users had gained what they saw as concessions. Part of the system had been removed, to some extent they were being listened to, and the system had been modified to work better. However, there were reactions to the ongoing implementation of the system.

Resistance to the ongoing implementation. As implementation proceeded, user reactions changed to "resist and modify." Many managers continued to use old manual systems such as the character cards, which were hand-written records of customer information including details of discussions with customers and any credit arrangements made. Several branch managers reported that the character cards had to be physically removed from the managers and filed away.

Due to inadequate training, staff tried to apply old procedures to the new system and did not use the functions that could make their work easier. Certain unusual transactions had not been included in the new system and the old system was left in place to conduct these. Staff discovered that the old system was still functional and then used whichever system suited their needs.

Critical mass and the subsequent reactions. "Critical mass" was reached when sufficient branches had been put onto the system to ensure that the others would accept the system in order to conform. At this point, users saw that the system could not be rejected so it made sense to cooperate. The reactions changed to acceptance, with minor resistance. The most usual reaction at this stage was that staff would only use a small range of functions essential to their work and would not explore possibilities.

5.2.3   Restoring the common sense of the organization

In the pilot implementation the system proved very difficult to work with and the managers soon became angry and resentful. The mood was such that normally compliant branch managers became outspoken. On several occasions the managers involved in the pilot implementation officially asked for the system to be taken out. As a manager put it: *"[T]he guys put their foot down and [said], either take the system and shove it or let us get involved and talk to us before you make changes."*

To some extent, this reaction was overdone. A branch manager at the time commented that: *"There were a lot of things that were blown out of all proportion."* Recollections of the events use terms such as *"I was rude and dogmatic"* and *"I went berserk."* This was unusual, as the managers were normally nervous to speak out about changes. A manager involved in the pilot implementation tells of an administrative manager who *"had a good chirp"* about the system and the next day phoned her manager to ask if she still had a job.

It took some time for the reasons for these outbursts to be uncovered. One reason was a sense of distrust toward the implementers. One branch manager graphically described how action would only ever ensue (his language was more colorful) if the managing director came to hear of the problems. If he was unaware of the problems, nothing would happen.

The management also saw the BDS as causing significant changes, particularly to their conditions of work as the new system was very different from the old, both in its interface and in its scope. In addition, there were technical problems, particularly excessive down-time and poor response times. As a result, concern was being felt about the lengths of customer queues at the tellers and the impact of the technical problems on customer satisfaction. Thus there was an underlying sense that the system would introduce uncertainty way beyond that which the organization could cope with.

The normally compliant staff also reported feeling anger and resentment toward the pilot implementation of the BDS, even though they were unable to express this at the time. However, the suggestions offered to the consultants that conducted the PPIS reflected their feelings. For example, the tellers complained that they were suffering tiredness and eyestrain due to the difficulty of using the long and complex menus. In addition, the informal network was quick to spread what one interviewee referred to as *"terrible stories"* about the system.

The response of senior management to these reactions was initially one of surprise. In retrospect, a senior manager reflected: *"I think we underestimated the initial resistance to change because it was just so different from the way in which we were doing it."* The surprise of senior management provoked them into an unusual situation. They decided to formally investigate the problems. This, in turn, prompted a mood of discomfort for the implementers. A senior manager in head office described this situation: *"The organization didn't like this report because we presented it up the line and it made people feel very twitchy and uncomfortable."*

The combined effect of resentment and anger from the managers and staff and the discomfort of the implementers provoked an unusual situation for the bank, a complete change in the approach to implementing systems. As a senior head office manager commented, this was a *"rude awakening."* As a consequence, the cashier part of the system was removed, work was commenced on improving the response times and ease of use, and the implementation strategy was changed. These efforts were welcomed by the managers and staff. It resulted in a mood that conceived the system as a possibility that could be dealt with.

Tyre and Orlikowski (1994) refer to a relatively short window of opportunity during which the users adapt work techniques and practices to the information system and, at the same time, adapt the system to the existing techniques and practices. After this period, there is a consolidation and little further change takes place. This process of adaptation in Bankco was considerably longer than their experience. Ongoing upgrades considerably modified the system over a period of two to three years. During this period, the managers and staff had to adapt and readapt to these changes.

As a result, for some time there was an underlying resistance to change and a reluctance to accept the BDS. Many of the managers' and staff's reactions were accompanied by the feeling that the BDS had disturbed the comfortable flow of everyday work. A particularly strong reaction was from managers who were uncomfortable with the computerization of many of the manual aspects of their work. This reduced their feeling of control as there was not the comfort of a tangible document to work with.

Once the implementation of the system settled down into a routine of 10 to 15 branches a month, the mood in which they came to see themselves changed to a sensing of possibilities. Typical reactions that were reported were:

> *[In the beginning] everybody came, very deflated and they just said, ugh, another day, another day with the system. But if you look at it now, it is actually a pleasure coming to work.*

> *It was, it was a bit scary because you...wonder[ed] now, gosh, are you going to be able to cope with your job now that it is all computerized. But as time goes by, you do, you learn...Now don't give me a job without a computer on my desk.*

This change in mood was reinforced by transfers that took place in the bank. Managers and staff who had moved to branches without the BDS discovered that they preferred the new system and would say so. Managers or staff who still opposed the system were left with the options of leaving the bank or transferring to other divisions. Several took this option. There is now a high level of enthusiasm for the system and most would complain if the company reverted to the old system.

6. Summary and Conclusions

Research streams in the field of information systems implementation have, in the past, been rather narrow (Kwon and Zmud 1987). This study seeks to build on the work of Myers (1994) which illustrates that the adoption and diffusion of an information system is a hermeneutic process that can incorporate conflict and opposition often with unexpected results.

This study is based on the understanding that, within the cohesive whole of the organization, the members have a capacity or ability to be. There is thus always the possibility that members will find opportunities or new ways to do things as well as the possibility that members will have particular moods (disposition or attunements) as a result of confronting these opportunities or new ways to do things. As a consequence of these moods, there is a questioning of the disruption. Organization disposition is the way of disclosing what matters to the members in this change situation. This way contributes to a tacit awareness of the technologies, strategies and events surrounding the adoption and diffusion of an information system and, subsequently, to a restoration of a common feeling of sense.

The concept of organizational disposition was incorporated into an integrative framework that provided the theoretical background for two case studies. In both cases, the reactions of the users brought about a dialectic process in which the moods experienced enabled the opening up or uncovering of the reasons for resistance and opportunities for response. The findings supported the view of Myers that neither technology nor the strategies of implementers can be viewed as the sole drivers of change.

The initial reactions of users were not consciously thought out but rather manifested themselves in moods as a tacit awareness of the (in)appropriateness of the change. Subsequently, the users attempted to articulate the underlying "reasons" for those moods. Even then it was not a rational process in which all the impacts on the organisation as a whole were assessed. Users tended to monitor their situation and sought to sustain some meaningful whole, "something that makes sense." In Bankco, the managers initially sought to reject the system rather than experience the discomfort of what they perceived to be an inadequate system. In Bevco, the decision to implement was delayed while the reasons for the initial discomfort were opened up and explored. The outcome was of benefit to the finance group but did not address the need of the regions. This supported the view of Markus and Keil (1994) that even a functional system will be rejected if the users are not motivated by to use it as intended.

The restoration of the common sense of the two organizations was reached through an ongoing process that integrated and reconciled the users' and implementers' dispositions—flowing from their "being-in-it-ness" (*Befindlichkeit*)—into a new coherent whole. This was a dialectic process in which the mutual adaptation of the sense of the whole were made by both users and implementers, *in* the situation, *as* it emerged. This supported Markus (1983) contention that resistance is not a problem to be solved so that a system can be installed as intended.

As suggested by Dreyfus (1991), the context of significance and the cultural structure that make up the cohesive whole of the organisation provide the background that lends intelligibility to the criticism and change that arise from the adoption and diffusion process (p. 161). The adoption and diffusion process impacts on the context of significance through the modification of activities and introduction of new "equipment" and new expectations for action. It is the tacit awareness of the (im)possibilities in the situation that is the most significant. Clearly the cultural horizon will make certain possibilities come up and others not. However, what is significant for adoption and diffusion is the common sense of the possibility to be in the emerging context.

Robey and Azevedo (1994) point out that Schein's (1990) definition of culture suggests that it is something that an organization has. They suggest that this risks reducing culture to a set of independent variables. They conclude that researchers should rather seek an understanding of the process through which the information system can either reinforce or revise cultures. It is this paper's claim that the notion of organization disposition can facilitate this understanding.

Finally, this study demonstrated that the adoption and diffusion of an information system is a disruption that breaks down the cohesive whole of the organization. Organizational disposition provides a way to uncover the holistic web of meanings and understandings which enable the members to resolve conflicts and to restore the cohesive whole. This restoration does not imply that the system would be successfully implemented. It may be that rejection of the system is the outcome necessary to restore the common sense of the organization, dysfunctional as it may seem to the implementers.

While this study contributes to an understanding of the adoption and diffusion process, its general application is limited. The conclusions were based on a limited set of data within two organizations. In addition, other factors within the organizations may have contributed to the outcome. Despite this, it is contended that the main conclusions of this study have not been invalidated. However, further work is needed, particularly on the role of dialogue in disclosing the moods and assertions of the users and implementers.

References

Berger, P. L., and Luckmann, T. *The Social Construction of Reality.* Middlsex, England: Harmondsworth, Penguin Books, 1966.

Clegg, S. R. *Frameworks of Power.* London: Sage Publications, 1989.

Coombs, R.; Knights, D.; and Willmott, H. C. "Culture, Control and Competition: Towards a Conceptual Framework for the Study of Information Technology in Organizations," *Organization Studies* (13:1), 1992, pp. 51-72.

DiMaggio, P. J., and Powell, W. W. "The Iron Cage Revisited: Institutional Isomorphism and Collective Rationality in Organizational Fields," *American Sociological Review* (48), 1983, pp. 147-160.

Dreyfus, H. L. *Being-in-the-world: A Commentary on Heidegger's Being and Time.* Cambridge, MA: The MIT Press, 1991.

Eisenhardt, K. M. "Building Theories from Case Study Research," *Academy of Management Review* (14:4), 1989, pp. 532-550.

El Sawy, O. A. "Implementation by Cultural Infusion: An Approach for Managing the Introduction of Information Technologies," *MIS Quarterly*, June 1985, pp. 131-140.

Fogarty, T. J. "Organizational Socialization in Accounting Firms: A Theoretical Framework and Agenda for Future Research," *Accounting, Organizations and Society* (17:2), 1992, pp. 129-149.

Frost, P. J., and Egri, C. P. "The Political Process of Innovation," in *Research in Organizational Behaviour* (Volume 13), L. L. Cummings and B. M. Staw (eds.). Greenwich, CT: JAI Press, 1991, pp. 229-295.

Giddens, A. *The Constitution of Society*, Cambridge, England: Polity Press, 1984.

Gouldner, A. W. "The Norm of Reciprocity: A Preliminary Statement," *American Sociological Review* (25:2), April 1960, pp. 161-178.

Haugeland, J. "Heidegger on Being a Person," *Noûs* (16:1), 1982, pp. 15-26.

Heidegger, M. *Being and Time*, tr. J. Macquarrie and E. Robinson. Oxford: Basil Blackwell Ltd, 1962.

Introna, L. D. *Management, Information and Power.* Hampshire, England: Macmillan, 1997.

Kofman, F., and Senge, P. "Communities of Commitments: The Heart of Learning Organizations," in *Learning Organizations: Developing Cultures for Tomorrow's Workplace,* S. Chawla and J. Renesch (eds.). Portland, OR: Productivity Press, 1995.

Kwon, T. H., and Zmud, R. W. "Unifying the Fragmented Models of Information Systems Implementation," in *Critical Issues in Information Systems Research*, R. J. Boland and R. Hirschheim (eds.). Chichester, England: John Wiley, 1987.

Markus, M. L. "Power, Politics and MIS Implementation," *Communications of the ACM* (26:6), 1983, pp. 430-444.

Markus, M. L., and Benjamin, R. I. "Change Agentry: The Next IS Frontier," *MIS Quarterly*, December 1996, pp. 385-480.

Markus, M. L., and Keil, M. "If We Build It, They Will Come: Designing Information Systems that People Want to Use," *Sloan Management Review*, Summer 1994, pp. 11-25.

Myers, M. D. "A Disaster for Everyone to See: An Interpretive Analysis of a Failed IS Project," *Accounting, Management and Information Technology* (4:4), 1994, pp. 185-201.

Newman, M., and Sabherwal, R. "Determinants of Commitments to Information Systems Development: A Longitudinal Investigation," *MIS Quarterly*, March 1996, pp. 23-54

Pettigrew, A. M. "Context and Action in the Transformation of the Firm," *Journal of Manage-ment Studies* (24:6), November 1987, pp. 649-670

Prescott, M. B., and Conger, S. A. "Information Technology Innovations: A Classification by IT Locus of Impact and Research Approach," *Data Base* (26:2/3), 1995, pp. 20-41.

Robey, D., and Azevedo, A. "Cultural Analysis of the Organizational Consequences of Information Technology," *Accounting, Management and Information Technology* (4:1), 1994, pp. 23-37.

Sahay, S., and Robey, D. "Organizational Context, Social Interpretation, and the Implementation and Consequences of Geographic Information Systems," *Accounting, Management and Information Technology* (6:4), 1996, pp. 55-282.

Schein, E. H. "Organizational Culture," *American Psychologist* (45:2), 1990, pp. 109-119.

Senge, P. M. *The Fifth Discipline.* New York: Currency Doubleday, 1990.

Silverman, D. *The Theory of Organizations.* London: Heinemann, 1970.

Tyre, M. J., and Orlikowski, W. J. "Windows of Opportunity: Temporal Patterns of Technologi-cal Adaptation in Organizations," *Organization Science* (5:1), February 1994, pp. 98-118.

Van Maanen, J., and Schein, E. H. "Toward a Theory of Organizational Socialization," *Research in Organizational Behavior*, Volume 1, L. L. Cummings and B. M. Staw (eds.). Greenwich, CT: JAI Press, 1979, pp. 209-264.

Walsham, G. *Interpreting Information Systems in Organisations.* Chichester, England: J. Wiley and Sons, 1993.

Wastell, D., and Newman, M. "Information System Design, Stress and Organisational Change in the Ambulance Services: A Tale of Two Cities," *Accounting, Management and Information Technology* (6:4), 1996, pp. 283-300.

Yin, R. K. *Case Study Research: Design and Methods*, 2nd ed. Thousand Oaks, CA: Sage Publications, 1994.

Zuboff, S. *In the Age of the Smart Machine: The Future of Work and Power.* New York: Basic Books, 1988.

About the Author

Brian O'Donovan is an Associate Professor at the University of the Witwatersrand and lectures in both information systems and management accounting. His Ph.D. drew on the philosophy of Heidegger to explore the dynamics of the adoption and diffusion of information systems in organizations. His current research interests and publication activities are concerned with the interaction between information systems and organizational change, strategic information systems, and control and performance measurement systems.

12 KEY ROLE PLAYERS IN THE INITIATION AND IMPLEMENTATION OF INTRANET TECHNOLOGY

Rens Scheepers
Aalborg University
Denmark

Abstract

Internet technologies have opened up vast possibilities for many organizations, including their application purely within the organizational boundary in the form of intranets. However, little is known about the organizational role players in the context of this technology. This paper describes a field study that examines the role players in the initiation and implementation of intranet technology in three large organizations in two countries. Using organizational innovation process theory and available intranet literature, the role players, their roles, their challenges and interrelationships are identified. Five key interrelated roles are isolated: the technology champion, organizational sponsor, intranet coordinator, intranet developer and content provider. Technology champions play the important role of initiating the technology in the organization. The organizational sponsor nurtures and protects the budding technology and its change agents throughout the process. The key role of intranet coordinator features prominently in organization-wide coordination, control and feedback across functional boundaries, while stimulating organizational use and content generation simultaneously. The intranet developer role is crucial for more advanced organizational application of the technology. Finally, the role of content providers is significant in creating a critical mass of content early during implementation to ensure progress.

Keywords: Intranet, initiation, implementation, role players, organizational innovation process.

1. Introduction

Internet technologies have opened up vast possibilities in terms of their organizational application. This is not only true for applications with an external focus (e.g., electronic commerce and "extranets" between organizations), but also for applications with an internal focus (e.g., organizational intranets). The latter technology is the focus in this paper.

An *intranet* is the application of Internet technology, more specifically World Wide Web technology, purely within the organizational boundary. Internet technology (web servers, browsers, etc.) is applied, but access is restricted exclusively to organizational members, for example, by firewalls (Oppliger 1997) or physically separating the intranet from external networks (firebreaks).

The organization's central information technology (IT) group has traditionally been the home of organization-wide information technology. Much of the literature concerning the introduction of organization-wide information technology concerns a paradigm where actors within this group play the key role in implementing technologies to address the needs of various user communities in the organization (e.g., Ginzberg 1981; Kling and Iacono 1984; Lucas, Ginzberg and Schultz 1990; Swanson 1987; Walsham 1993).

Early published research into implementation of intranet technology seems to indicate a shift in this paradigm. Some intranet studies have found that the technology was in fact initiated by decentralized actors outside the IT group (Bhattacherjee 1998; Jarvenpaa and Ives 1996). Lyytinen, Rose and Welke (1998) stress the profound impact of the continuing evolution of Internet (and intranet) computing infrastructure on organizational processes. They predict that the ubiquitous nature of these technologies will blur traditional IS roles (e.g., users will become developers); demand new roles as technologies and media coalesce; and further that these changes will be on a much grander scale than ever before (e.g., as in the case of end user computing).

These early research efforts lead us to expect a new "cast of characters" who may be involved in the initiation and implementation of intranet technology. This paper, therefore, seeks to answer the following questions: who are the key intranet role players and what roles do they play? Moreover, what are the key challenges associated with these roles and how do the roles interrelate?

The paper is structured as follows: The theory of the organizational innovation process is drawn upon to identify key role players. These roles are then examined in the context of available intranet literature. A report on a field study of three intranet cases is made and these cases are analyzed using the innovation process theory and intranet literature. The intranet role players are identified and each role's key challenges and interrelationships are described. Finally, the findings are discussed, conclusions are drawn, and issues for future research indicated.

2. Role Players in the Organizational Innovation Process

The initiation and implementation of a new technology within an organization can be regarded as an organizational innovation process (Zaltman, Duncan and Holbek 1973).

Rogers (1995) describes the organizational innovation process as consisting of two broad activities: initiation and implementation.

Initiation includes the information gathering, conceptualizing, and planning for the adoption of an innovation, leading up to the actual adoption decision. Implementation includes all the events, actions and decisions involved in putting the innovation into use (Rogers 1995, p. 392). In general, a new innovation is initiated and adopted by a small group of actors. However, the implementation of the innovation concerns the communication of the innovation to a much larger group of organizational members who should also adopt and ultimately put the innovation into use (voluntary or involuntary) (Wynekoop and Senn 1992).

A number of role players are recognized in the context of the organizational innovation process. Initiation of an innovation is usually attributed to the efforts of innovators, and often an individual innovator (Schön 1963). In some cases, the innovator actually works outside the formal research teams of organizations, and as a consequence without organizational support. Schön notes that due to the considerable resources needed for organization-wide implementation of an innovation, someone with the required power and prestige in the organization needs to emerge to take control of the innovation, else the process stalls. This role is played by an organizational sponsor (Humphrey 1989; Rogers 1995). Although the sponsor role is also labeled a "champion" by some authors, the term "organizational sponsor" is used here to avoid confusion later on when referring to the literature describing roles in information technology innovations. Once an organizational sponsor has emerged for the innovation, the innovation process can enter the implementation stage. Change agents play an important role when the implementation of an innovation demands changes in the behavior of organizational members. Change agents communicate the innovation to organizational members and seek to influence their behavior with respect to the innovation (Rogers 1995).

In the context of information technology (IT) innovations, the roles of "technology champion" (innovator), organizational sponsor and change agents are also widely recognized. A technology champion is an organizational member who initiates the use of a new information technology in the organization (Beath 1991). Technology champions are sometimes informal experts whose main responsibilities are "doing work" and not to experiment with new technology (Attewell 1992). In many cases, technology champions lack the necessary authority and/or formal resources (Beath 1991) and thus seek to draw the attention of senior management toward the technology (Lawless and Price 1992). A sponsor is usually a senior manager who realizes the potential of the information technology in the organization. She either allocates organizational resources towards further implementation of the technology herself and/or negotiates such resources with her senior colleagues (Martinsons 1993). Once the sponsor has taken "ownership" of the technology, the role of the technology champion disappears into the background (Orlikowski et al. 1995). Beatty and Gordon (1991) describe the role of the organizational sponsor in subsequent implementation stages as that of a "godfather" who empowers and supports agents of this change in the organization.

Implementation of an innovation is an organizational change process and concerns not only the technical, but also the political and cultural dimensions of the change (Keen 1981; Markus 1983; Walsham 1993). Change agents come into play to effect the implementation of a new information technology (Beatty and Gordon 1991; Markus and

Benjamin 1996; Orlikowski et al. 1995). These agents play a key role in spreading the new information technology to organizational actors. Change agents should be both technically and politically astute and capable of addressing the problems associated with the implementation. Markus and Benjamin describe three models of the change agent role: the traditional model, the facilitator and the advocate model. Simply stated, the traditional model assumes technology causes the change and the change agent has a limited role beyond building the technology. In the facilitator model, the change agent tries to reduce friction between users and technology specialists, thereby enabling better IT management and use. In the advocate model, the change agent attempts to influence people's behavior in particular directions that the change agent views as desirable. Markus and Benjamin note that a likely change agent role description includes elements of all three models.

3. Available Intranet Literature

Probably due to the attention in the computer press, intranets have been a "hot" topic and a number of "popular" books have resulted (e.g., Bernard 1996; Dyson, Coleman and Gilbert 1997; Hills 1997). These mostly describe the marvels of intranet technology, how it integrates text, graphics, sound, and video, and the "ease" with which attractive intranet websites can be built. However, in terms of providing insight into intranet implementation, none of these sources go much beyond anecdotal "success stories."

A few published research studies highlight some of the general challenges associated with intranet technology implementation in the organizational context (e.g., Bansler et al. 1999; Goles and Hirschheim 1997; Scheepers and Damsgaard 1997).

Damsgaard and Scheepers (1999a) describe intranet technology as multi-purpose, richly networked and malleable in terms of its application. They isolate a number of intranet technology "use modes." These range from simple uses such as the *publishing* of home pages, newsletters, technical documents, product catalogues, employee directories, etc., to more advanced uses such as organization-wide *searching* for information; *transacting* with functionality on intranet pages and other organizational computer-based information systems (e.g., legacy systems); *interacting* between individuals and groups in the organization (e.g., via discussion groups, collaborative applications); and possibly even *recording* the computer-based "organizational memory" (e.g., best practices, business processes). The authors note that unlike many interactive media (Markus 1987) where it is sufficient to attract only a critical mass of users (e.g., e-mail), intranets require a critical mass of both users and content to be pervasive. Intranet implementers are thus faced with a double problem: attracting users and getting them to generate sufficient content. A further complexity is that various levels of intranets can co-exist in the organization, because the technology can be implemented centrally in the organization (as a corporate intranet), but decentralized actors (e.g., in divisions, departments or functional groups) can simultaneously implement "child-intranets."

Damsgaard and Scheepers (1999b) warn that three main challenges must be overcome to ensure eventual institutionalization of an intranet, else the technology is bound to stagnate. First, if a sponsor does not nurture the technology it cannot evolve beyond its emergent beginnings. Second, if a critical mass of both users and content cannot be reached simultaneously, the intranet will not progress. Finally, if the intranet remains uncontrolled (i.e., "grows wild"), it will be perceived to be useless and therefore users will abandon it.

A few studies refer peripherally to other role players during the initiation or later during implementation of intranet technology, but none of these studies go into much depth regarding these roles.

Jarvenpaa and Ives found that the introduction of Web technology was neither initiated by the organization's IT group or by senior management. In the organizations they studied, the technology was initiated by decentralized technology champions from a variety of organizational functions outside the IT group. These champions cooperated as "virtual teams" of "volunteers." The IT groups of the organizations eventually became involved with the technology, but very much in a reactive fashion.

Romm and Wong (1998) describe the implementation of an intranet in an Australian university setting. In contrast to the findings of the Jarvenpaa and Ives research, in their study the intranet was led by the IT department, with a high involvement of senior management. An intranet committee was formed and the project coordinator attempted to convince university departments to add their information to the intranet. Despite his efforts, the intranet was unsuccessful because departments did not live up to their promises of converting and adding content to the intranet.

Bhattacherjee reports on a case study of the "Global Village" project, which describes the implementation of the intranet at US West Communications. The intranet project was initiated by the Finance organization within US West. The Vice President of Finance acted as organizational sponsor and hired a project manager who was a "technological visionary" to implement the intranet. Together with staff within the Finance group, these actors comprised the intranet project group and successfully managed to establish company-wide interest for the intranet. The study reports that "ownership" of intranet content was transferred to individuals in various user departments of the organization.

In a speculative paper, Lyytinen, Rose and Welke predict that the ubiquitous nature of Internet technologies will blur traditional IT roles (e.g., users will become developers) and that we will see new organizational roles as technologies and media design coalesce. They foresee roles of artists and content creators becoming increasingly important and highlight the role of system developers who should be even more capable of cooperating in multi-skilled teams.

4. Field Study

The case study research approach is especially appropriate in new topic areas (Eisenhardt 1989). The field study reported here was based on the explorative multi-site longitudinal case study approach (Yin 1989).

Empirical research was conducted in the head office environments of two large organizations in South Africa and one in Denmark. In South Africa, the corporate intranets of the CSIR (www.csir.co.za), a large semi-government R&D organization, and that of Telkom (www.telkom.co.za), the national telecommunications provider, were studied. In Denmark, the intranet at the headquarters of the LEGO Group (www.lego.com), a global toy manufacturer, was studied. Qualitative data was collected from a variety of actors who were involved in the implementation of the technology in these organizations (principle of multiple interpretations, Klein and Myers 1999).

After gaining initial access to the organization, the intranet role players were identified on a peer-referral basis (Howell and Higgins 1990) and interviewed. The aim was to isolate intranet "roles" (Rogers 1995) rather than to identify specific individuals involved in the intranet. Given the duration of this research, this approach allowed us to overcome the problems of variation in position descriptions, new appointments, resignations, etc.

The study aimed to use a flexible data gathering strategy and to find a representative set of data (Benbasat, Goldstein, and Mead 1987). Semi-structured interviews were the primary data collection method used. Open-ended questions were organized into an initial questionnaire using theoretical constructs from the literature and similar research efforts. The initial questionnaire was used in a pilot study and was subsequently refined to improve understandability and comprehension. The refined questionnaire was used for the main data gathering.

The interview guide was used as a basis, but the interview was allowed to digress if other interesting issues surfaced. Interviews also covered normal background information about the size and type of business and about the affiliations and education of the interviewee, etc. All interviews were tape-recorded and notes were taken during the interviews. Other data were also collected, including notes from informal discussions, e-mails, policies, reports, demonstrations, own inspection of the intranet, and even promotional material that was used during some implementations. Interviews were transcribed and shared with the interviewees to check for possible errors and omissions and to evaluate the validity of the interpretation of their "story" (Klein and Myers 1999). Based on all the data and transcripts, rich descriptions of the roles of the various role players were obtained at each site.

Table 1 summarizes the research duration and number of interviews at each site. In two of the cases, apart from an initial "baseline" round of interviews, follow-up interviews were done, mostly with the same interviewees.

Table 1. Summary of Data Collection

Organization	Research duration	Number of Interviews
CSIR Corporate, South Africa	September 1997 (pilot); December 1997-January 1998 (baseline); November 1998 (follow-up)	3 (pilot) 8 (baseline) 8 (follow-up)
Telkom Head Office, South Africa	December 1997-January 1998 (baseline); November 1998 (follow-up)	5 (baseline) 4 (follow-up)
LEGO Group Headquarters, Denmark	August 1998 and October 1998 (baseline)	8 (baseline)

5. Three Cases of Intranet Initiation and Implementation

5.1 The CSIR IntraWEB

The CSIR in South Africa is Africa's largest scientific and technological research, development and implementation organization. It is semi-governmental and does industrial contract research in the public and private sectors in specialized technological areas. The CSIR consists of nine semi-autonomous divisions that report to a president at the head office. In September 1996, the organization started to implement a corporate intranet called the CSIR IntraWEB.

Prior to the implementation of the corporate intranet, computer scientists in a few of the divisions started independent intranet experiments within their environments. Their enthusiasm spread informally to more of the divisions and also to the organization's central Computing Services group. Technical staff in the Computing Services group started constructing an experimental organizational intranet home page. The home page effort gained the support of managers in the Computing Services group and a project leader was appointed within the Computing Services group to take the idea further. The intranet project leader headed up a small team of programmers.

These efforts came to the attention of the CSIR's Vice President of Policy and Technology. The Vice President realized the organization-wide potential of the technology and played a key role in approving funding toward the initiative. The Vice President appointed a Quality and Systems manager with one of her tasks to coordinate the CSIR intranet in conjunction with the intranet project leader at Computing Services. She commented on her role in the implementation:

> *The Vice President sees me as a custodian…not to do the work, but to coordinate and see that it is done. In other words, to manage the resources.*

Development of the intranet progressed further in the organization and it became known as the CSIR IntraWEB. The Quality and Systems manager formed a steering group consisting of divisional intranet representatives and the intranet project leader. Together with the steering group, she initiated a number of organization-wide initiatives to further awareness of the intranet. This included a major organizational campaign with intranet posters being put up throughout the CSIR, giving away coffee cup "coasters" with the intranet information on them, intranet talks and an intranet "treasure hunt" with prizes to be won. The Vice President remained firmly supportive throughout these efforts, also in convincing his executive colleagues of the organizational advantages of the technology.

The Quality and Systems manager initiated a project via which divisional and corporate business plans, policies, management models and quality plans were put on the intranet. Given the semi-autonomous nature of the divisions, she mentioned this required considerable "lobbying" of some divisional managers.

The intranet project leader initiated a number of developments to integrate existing computer-based systems in the CSIR with the intranet. These included "web-enabling" many of the existing information systems in the CSIR to allow access from the intranet and also new intranet applications such an employee directory, which would combine data from a number of systems (e.g., human resources, security access system, etc.). In many cases, this necessitated the project leader to intervene to convince staff in other corporate units to tailor their developments and processes to allow for integration with the intranet. The project leader also initiated the implementation of an intranet search engine to allow organization-wide searches.

Follow-up research after 10 months revealed that after the initial intranet efforts, many divisions and corporate units had subsequently designated specific staff members to take responsibility for the content of their local intranet sites. This created the need for intranet design courses, which were set up by the Quality and Systems manager. Following these courses, a staff member located in the Communications unit created a fairly sophisticated intranet site that automated and improved the previous process by which staff members ordered corporate gifts for their clients. The editor of the staff magazine started to publish the magazine also via the intranet and she was trying to persuade her superior to suspend paper publication altogether.

The intranet project leader held responsibility mainly for the corporate part of the intranet and for the integration of even more organizational systems with the intranet. He described his role further:

> *[W]e are moving towards a development model where my team and I will be responsible for intranet project management. Some develop-ment can take place in the divisions, but we will ensure a coordinated approach.*

The Quality and Systems manager retired, but the Vice President transferred her role to a senior manager in the newly created CIO office. The CIO manager established a CSIR "Web Council" with senior representatives from all divisions to decide on issues such as content standards, use policies and future directions of the CSIR's intranet and Internet applications.

5.2 Telkom Intranet

Telkom South Africa is a government owned company that provides around 40% of all telephone services on the African continent. The organization consists of a head office, service groups and regional offices in all the main centers of southern Africa. Within Telkom, several individuals in service groups (mainly IT-related environments) and the regional offices were "playing" with pilot intranet sites. These started spreading through informal networks and a large number of "island" sites resulted. A regional technician who created one such pilot site commented:

> *I'm not "officially" supposed to be involved with the intranet. I just started using it to make my job easier. They can say to me tomorrow to leave it and focus on what I'm supposed to do.*

He added:

> *People are joking with me, but I put my name on all the intranet pages I create. I want to go further and get a post as a programmer so I can have a budget for software.*

These efforts prompted Telkom's Information Executive to establish a small team of four staff members within the IT Services Group and tasked them with the responsibility of integrating the "island" intranets into a cohesive whole. This marked the "formal" start of Telkom's corporate intranet (Damsgaard and Scheepers 1999a) in February 1997.

The intranet team consisted of an information specialist, a graphics designer, a web developer and a network security expert. The team implemented a basic corporate intranet and hyperlinked the unit level intranets to it. The main feature of the corporate intranet was structured "content book" pages of information topics with hyperlinks to the intranet sites throughout Telkom. The team structured the topics and maintained the hyperlinks on an ongoing basis. The "content book" became the primary mechanism for locating information on the intranet (no central search engine existed). Over time, forums for interaction on a variety of organizational issues were added. Many intranet sites, however, still contained IT related information.

Although the Information Executive actively supported the implementation, not all of his senior colleagues were sold on the concept initially. The Information Specialist commented:

> *At the moment top management is a little bit weary of this new thing. They've got this idea that surfing the intranet means playing around all day. "You are going to surf pornographic sites, you are going to play games." We have to get it across to them that it is of business value to them. We are not at the point where management is giving it their full support.*

The Information Executive commissioned a draft intranet policy for Telkom. The policy recommended, among other things, "proper use" of the technology and content standards.

Apart from playing the main role in structuring the content of the corporate intranet, the Information Specialist embarked on a variety of tactics to further intranet use. She featured a monthly "best intranet site" on the corporate intranet; she initiated the distribution of information about the intranet along with employees' pay slips and she launched an intranet poster campaign. As a result, many organizational units started to develop their own intranet presence. These ranged from sites that were very rich in information content to some which a senior manager described as merely "this is me and this is my team" sites.

Follow-up research after 10 months revealed no significant new developments. The Information Executive largely withdrew from the intranet initiative due to organizational demands for his leadership in addressing the Year 2000 problem. Use of the intranet was still mainly for publication and interaction, although the need for more advanced uses, especially transacting with organizational databases, was identified. The task of maintaining the content of intranet sites was seen as an "add-on" to staff members' existing jobs. The Information Specialist started to quality assure the content of intranet sites. In a few cases, she removed links from the "content book" to sites which contained outdated information or did not meet the content standards.

5.3 LEGO Web

The international LEGO Group, in which the original company was established in 1932, is one of the world's largest toy manufacturers. The LEGO Group is owned and managed by the Kirk Kristiansen family in Denmark. The organization provides creative experiences, construction toys, educational materials, lifestyle products, family parks and media products for children all over the world.

Following the development of the LEGO Group's Internet site, the programmer who was involved in the project, in conjunction with his superior, initiated the idea of an intranet to senior management. A director of the LEGO Group approved a one-man project (the same programmer) and an intranet demonstration prototype was built. The prototype was demonstrated to top management who approved further development. The demonstration happened to coincide with the earlier launch of a new management strategy in the LEGO Group that aimed, among other things, to improve interdepartmental communication. The potential of the intranet technology supporting this management strategy was quickly realized.

Implementation of the envisioned intranet (called LEGO Web) started toward the end of 1996 and early 1997. With strong support from the LEGO Group director, a small team of four programmers was established within the IT group. The LEGO Web project leader headed this team. The Director further appointed a Web editor in the Information and Public Relations department. The intranet project leader and the Web editor cooperated closely. Throughout the implementation, the director ensured adequate funding was available for intranet developments.

Considerable intranet functionality was developed by the intranet project team. This included organizational wide search facilities, a number of organization-wide intranet applications (e.g., a global employee directory integrating e-mail, phone numbers, picture of employee) and facilities for interaction (e.g., intranet discussion groups). Although direct transacting with existing systems was not possible at the time of this research, functionality was built to "upload" information from various systems (e.g., manufacturing data, product information). This allowed for high quality graphical content (e.g., pictures of products and packaging).

Once the basic intranet infrastructure was in place, a decision was made to formally designate about 80 employees worldwide as dedicated LEGO Web "content providers." The Corporate IT Manager reflected on this decision:

> *[W]e are talking about the "hen and the egg" problem. We had to get critical mass here, otherwise people would say "this is nice, but there's nothing on it."*

The Web editor controlled the addition of new content to the intranet. She formulated content standards for intranet pages and assured the quality of all intranet content received from content providers prior to its addition to the intranet. The Web editor supported the content providers (e.g. in terms of training and the use of content standards).

The Web editor promoted the intranet via a number of articles in the staff newsletter. There was also a formal launch of the intranet attended by the CEO of the LEGO Group. The Web editor engaged in interdepartmental information exchanges such as to convince department A that they should provide content on the intranet for department B to use. This proved difficult in some cases and she complained of a lack of middle management participation and a tendency that some departments were advertising themselves rather than providing useful information.

The development of the organization-wide employee directory required telephone data to be available more regularly and in an electronic format. This required process changes in the department responsible for the previously printed version of the telephone directory. The intranet project leader intervened to convince this department to convert their paper process to an electronic process so this data could be used on the intranet. Reflecting on her responsibilities, the intranet project leader mentioned she struggled to deliver on existing requirements while "keeping pace" with the rapid new developments in intranet technology.

The newly designated content providers were pleased with their new responsibilities and saw their new task as an upgrade of their work. This was echoed in the following statement:

> As secretary, you usually just type what other people think. Now all of a sudden I have an identity of my own. I have got the right to contact [people] with information, instead of going through somebody else.

Though the intranet was well supported by management, some content providers needed to change old ways of doing things in their local units:

> We were working with paper before. Here we have a good opportunity to make it cheaper for the company. The user can always be sure they are working with the right version and they can have direct access to it. In my environment, I'm the one that's pushing the intranet.

6. Analysis of the Cases

Following the principle of abstraction and generalization (Klein and Myers 1999; Walsham 1993), the theory of the organizational innovation process as well as the available intranet literature are drawn upon and a number of abstract roles in the initiation and implementation of the technology are identified and analyzed. A summary of the identified roles in terms of role descriptions, key challenges and interrelationships appears in Table 2.

Table 2. Summary of Roles, Descriptions, Challenges and Interrelationships

Role	Role description	Key challenges	Interrelationships
Technology champion	"Reinvents" technology in organization, establishing the intranet concept	"Selling" the intranet concept to senior management	Organizational sponsor
Organizational sponsor	Negotiates organizational resources; Supports and nurtures organization-wide implementation of technology; Approves usage policies and content standards; "Godfather" for intranet agents	Owning the responsibility for the intranet and associated political risks	"Grabs" intranet concept from technology champion; Supports intranet coordinator, developers
Intranet Coordinator	Organizational wide intranet change agent, manages cross-functional intranet initiatives; Advocates intranet; Information "broker" between units; Manages content quality	"Double problem" of getting critical mass of both users and content while ensuring content quality	Dependent on support of sponsor; Strong relationship with intranet developer
Intranet Developer	Leads technical intranet infrastructure development; Change agent where intranet's technical aspects demand new or changed systems and organizational processes	Keeping pace with new technological developments, while delivering intranet infrastructure	Strong relationship with intranet coordinator, but "second fiddle"; Dependent on support of sponsor
Content Provider	Local change agent within own unit; Campaigns for use of intranet in departmental processes; Creates, maintains and translates content on behalf of others in unit	Designing appropriate content; Resistance to change at local unit level	Dependent on support of intranet coordinator; Dependent on indirect support of the sponsor

6.1 Technology Champions

The role played by intranet technology champions can be identified in all three cases. In the LEGO Web case, the intranet technology champion was an individual programmer who championed the intranet concept based on his earlier involvement in the development of the organization's Internet site. In the CSIR case, the various divisional computer scientists were the technology champions. In the Telkom case, the technology champion role was played by regional technicians and also IS staff in various head office units. In both these cases, the champions interacted and learned from each other's independent experiments via informal peer interaction.

Even though the same technology which is used on the Internet is also used for intranets, technology champions "reinvented" the technology (Attewell 1992) and established the intranet concept in the organizations. The technology champion(s) would then attempt to "sell" the concept to senior management through pilot applications (LEGO Web and Telkom cases) which could demonstrate the technology's potential.

In all cases, the efforts of the technology champions were unsolicited. In no case was senior management found to have initiated the intranet. However, it seems in all cases that the champions had the freedom to experiment with the technology. The motives of intranet technology champions seem to vary (e.g., curiosity, even self-interest).

In the LEGO Web case, the technology champion was actually within the IT group (as found by Romm and Wong), but as the other two cases demonstrate, technology champions operated independently and outside of the IT group (as found by Bhattacherjee and by Jarvenpaa and Ives).

As predicted by Orlikowski et al., in all of the cases, the technology champion role disappeared into the background once an organizational sponsor and subsequent implementation agents came onto the scene. However, as the Telkom case shows, many of the original IT content sites remained active and it seems some technology champions switched roles afterward and became content providers.

6.2 Organizational Sponsor

In all three cases, an organizational sponsor emerged to "take ownership" of the intranet technology. In the CSIR case, the sponsor was the Vice President of Policy and Technology. At Telkom, the sponsor was the Information Executive, while in the LEGO Web case the Group Director played this role.

In all of the cases, an organizational sponsor effectively "grabbed" the intranet concept from the technology champion(s) (Damsgaard and Scheepers 1999b). Although in each of the cases a number of decentralized "child-intranets" co-existed, the intranet sponsor role was invariably played by a single key individual in the organization. In all cases, the sponsor was a senior manager who realized the technology's organizational potential (based on the "selling" efforts of the technology champion(s)). As the Telkom case demonstrated, not all senior managers were sold on the intranet concept initially. This means that the role of the sponsor can involve political risk-taking.

In all cases, the sponsor played an active role in allocating or negotiating funding and resources toward intranet implementation. The sponsor also engaged change agents (such as intranet coordinators and developers) to lead organizational implementation of the intranet. It is interesting to note that, in all three cases, the sponsor personally appointed the intranet coordinator.

In subsequent phases, the sponsor did in fact play the "godfather" role (as described by Beatty and Gordon), especially in terms of supporting the intranet agents (CSIR, LEGO Web cases). The sponsor was also instrumental in approving use policies and content standards (LEGO Web and Telkom cases). As the Telkom case shows, a problem may result if the sponsor prematurely withdraws his sponsorship. This may mean that the agents depending on his support may well be left vulnerable (as found by Beatty and Gordon).

6.3 Intranet Coordinator

The prominent role of intranet coordinator is identified in all three cases. The intranet coordinator played a central role in the implementation process and was invariably highly visible, but strongly dependent on support of the sponsor. In the CSIR case, the manager of Quality and Systems played this role initially and the role was later continued by the manager in the CIO office. In the Telkom case, the Information Specialist in the IT group played the role, while in the LEGO Web case, the Web editor located in the Information and Public Relations Department played the role. In all cases, the intranet coordinator was located in a central position in the organization.

The cases describe the intranet coordinator as a powerful change agent and the role has strong facilitating and advocating facets (as described by Markus and Benjamin). The focus of the coordinator's role is organization-wide, transcending functional boundaries and the "integrating mechanism" of various intranet activities in the organization. The coordinator was observed to be very much dependent on the organizational sponsor for political support (as in other studies, e.g., Beatty and Gordon). However, in the present cases, the coordinator role was found to be much more prominent than in earlier studies (e.g., Romm and Wong).

The key challenge of the intranet coordinator is to influence the behavior of various organizational actors to change with respect to the new technology. This meant positioning the intranet as a new medium to carry organizational information and making organizational actors with a need for such information aware of the intranet, thereby addressing the "double problem" of reaching a critical mass of both users and content simultaneously (Damsgaard and Scheepers 1999a). In terms of advocating intranet use, various approaches by the coordinator were seen in the cases, e.g., organization-wide campaigns, presentations, articles in staff newspapers, and intranet treasure hunts. In the CSIR and LEGO Web cases, in order to ensure a critical mass of useful content, the coordinator was seen to facilitate cross-functional, intranet initiatives such as the creation of an organization-wide employee directory (in conjunction with the intranet developer). She would support intranet content providers and lobby middle managers (LEGO Web case). In the CSIR and LEGO Web cases, the coordinator was seen as intervening as an "information broker" between departments, convincing them to exchange information via the intranet rather than via traditional means. The coordinator also played a central role in formulating use policies, information standards, and quality assurance of content (Telkom, LEGO Web cases).

6.4 Intranet Developer

In the cases, the role of an intranet developer can also be identified. In the CSIR case, the role was played by the Intranet Project Leader and, in the LEGO case, the role was played by the LEGO Web Project Leader. In the Telkom case, the role of intranet developer as defined here was largely absent.

The role of the intranet developer encompasses the main responsibility for the technical intranet infrastructure developments. The intranet developer is also a powerful change agent who has the ability to intervene where the technical intranet infrastructure requires new systems, changes to existing systems or organizational processes (CSIR, LEGO Web cases). In the CSIR and LEGO Web cases, this role was typically played by a senior project manager within the organization's corporate IT group, but with political backing by the organizational sponsor.

Although in some cases the organization's corporate IT group "reacted" to the intranet initiation by outside technology champions, the intranet developer role later became crucial in leading developments associated with the more advanced technology uses such as searching (e.g., implementing organizational search engines) and transacting (building links to "legacy" systems and integrating existing organizational databases with the intranet).

As witnessed in the LEGO Web case, the intranet developer role also has aspects of a change agent. In this case, the developer intervened to get departments to alter their departmental processes and systems to allow the creation of organization-wide intranet applications (such as the organizational employee directory). In the CSIR and LEGO Web cases, the intranet developer worked in close collaboration with the intranet coordinator, and was also strongly dependent on the sponsor for political support and funding for developments.

The absence of the intranet developer role in Telkom explains the limited use of the technology (only publication and interaction). No organizational search facilities were available, and the intranet coordinator had to resort to "simulating" this functionality manually via the "content book." The need for the developer role was, however, realized in this case.

Although in the CSIR and LEGO Web cases small teams accomplished intranet technical developments, the key role played by the intranet developer personally was well recognized by interviewees. However, as compared to the highly visible role of the intranet coordinator, the role of the intranet developer can be seen as the "second fiddle."

The LEGO Web case demonstrates the impact of rapid new developments in the technology on the intranet developer role, where the interviewee described her struggle between delivering on requirements and "keeping pace" with new technological developments.

6.5 Content Providers

The role played by content providers (Bhattacherjee 1998) was identified in all cases. In the Telkom case, these roles were not formalized. In the LEGO Web case, the content provider role was formalized up-front, while formalization of the role happened later in the CSIR case.

Table 3. Comparison of Intranet Roles Across Cases

Role	CSIR IntraWEB	Telkom Intranet	LEGO Web
Technology champion	Role played by independent decentralized actors during initiation	Role played by independent decentralized actors during initiation	Role played by individual actor during initiation
Organizational sponsor	Role played throughout initiation and implementation	Role played initially, withdraws during implementation	Role played throughout initiation and implementation
Intranet coordinator	Role prominent throughout implementation	Role prominent throughout implementation	Role prominent throughout implementation
Intranet developer	Role present throughout implementation	Role largely absent	Role present throughout implementation
Content provider	Role initially informal, later formalized	Role remained informal throughout implementation	Role formalized very early in implementation

The content provider plays a role of creating and maintaining new intranet content and converting existing information to the intranet at the local unit level, usually on behalf of other colleagues in the unit. However, as is evident in the LEGO Web and CSIR cases, this role extends beyond the pure technology dimension only (Markus and Benjamin 1996). In these cases, the finding was that content providers would operate as local change agents within their own units, advocating more use of the intranet in local processes, and trying to influence their colleagues' behavior.

In the CSIR case, a content provider was observed initiating a fairly advanced process automating the ordering of corporate gifts. Here is an example where the technology blurs roles, because a normal office worker with minimal training was able to develop the kind of functionality one may expect from a "systems developer" (Lyytinen, Rose and Welke 1998). Furthermore, as predicted by Lyytinen, Rose and Welke, in the Telkom and CSIR cases, the intranet technology was seen as creating a need for content providers with artistic or design skills.

As is evident from the Telkom and LEGO Web cases, the key challenges the content provider faces are the design of appropriate content and dealing with their colleagues' resistance to change. In this regard, content providers are dependent on the intranet coordinator for direct support, but also dependent on "indirect" support from the organizational sponsor.

In the LEGO Web case, the comment from the content provider about gaining her "own identity" gives us a rare glimpse at the often intangible potential of the technology in "empowering" workers (Clement 1994).

7. Discussion

The discussion is divided in two parts. First, the power of organizational innovation process theory in identifying the intranet role players is reflected upon. Second, the significance of the identified roles is discussed and compared across the cases.

Using the theory of the organizational innovation process with its associated role players has been a useful "theoretical lens" to identify role players in the intranet context. In each case, the roles predicted by the theory could easily be found and the way in which these roles manifested during initiation and implementation could be examined. The theory did, however, limit the study to only identifying roles during implementation where intranet agents attempted to influence, change and stabilize new behavior patterns of fellow actors in the organization. The theory would be inappropriate to identify roles in subsequent routinized use stages where there is less emphasis on "change."

The analysis indicated that, in the case of intranet technology as an interactive medium, the clear demarcation between the "implementation" stage and subsequent "use" stage (Lucas, Ginzberg and Schultz 1990) becomes blurred. Since the interactive medium's implementation success depends on a critical mass of early content to draw users, this moves the role of "users" such as content providers to within the implementation stage itself.

The roles identified here each have their own specific significance. A cross-case comparison of the roles appears in Table 3.

Technology champions played the important role of initiating the technology in the organizations. As also found by Jarvenpaa and Ives, it was noted that the senior management did not commission the intranet in any of these cases. It was the unsolicited efforts of technology champions that prompted implementation of the technology.

The role of organizational sponsor was significant in all of the cases. Since senior management "did not ask" for the intranet, this role was found to be especially significant in nurturing and protecting the budding technology and its agents throughout the process. Withdrawal of the sponsor during the implementation process seemed to jeopardize the implementation process in one case.

The role of intranet coordinator featured prominently in all of the cases. The role was significant in the sense of providing an organizational wide coordination, control (Ashby 1956) and feedback mechanism (Beer 1959) across functional boundaries while addressing the double problem of stimulating organizational use and content generation simultaneously.

The role of intranet developer was significant when more advanced organizational application of the technology was needed. The absence of this role in one case was seen as restricting the organization to only elementary use modes.

Finally, the role played by content providers seems significant in creating a critical mass of content early in the process to ensure progress. It was interesting to compare the effect of different approaches in terms of when (and if) the organizations decided to formalize this role. It seems that more accelerated intranet implementation depends on early formalization of this role.

Some of the intranet literature has highlighted the significance of intranet steering groups and teams (e.g., Jarvenpaa and Ives 1996; Romm and Wong 1998). However, as found in the present cases, although there was mention of steering groups and intranet teams, the individual role players were the main driving forces in the process.

8. Conclusion

The conclusion is that there are five key interrelated roles in the initiation and implementation of intranet technology. As predicted by the innovation process theory and intranet literature, the role of technology champion and organizational sponsor during intranet initiation are identified. During implementation, the technology champion role disappears into the background. The organizational sponsor takes ownership of the technology and empowers and supports three specific change agent roles: the intranet coordinator, intranet developer and content provider.

As predicted by Lyytinen, Rose and Welke, evidence was found that the traditional IS roles of "user" and "developer" are indeed blurred in the case of intranet technology. Some intranet content providers can be seen as "users," and also as "developers" of content and functionality.

Evidence was found that, in the case of intranet technology initiation and implementation, some limelight is stolen from the organization's central IT group (also observed by Jarvenpaa and Ives and by Bhattacherjee). Leading roles were played by technology champions, the organizational sponsor, the intranet coordinator and content providers who often resided outside the central IT group. However, it is concluded that the key role of intranet developer (mostly located within the IT group) remains crucial for more advanced organizational application of the technology.

The study has a number of limitations. First, only the initiation and implementation of intranet technology in large, established, and hierarchical organizations were examined. The roles identified and explored here may manifest themselves differently in other settings. For example, in small or medium sized enterprises, one may find that different roles may need to be played by the same individual actor. Furthermore, the study was limited in the sense that it only focused on the initiation and implementation stages. The data did not allow examination of subsequent stages of routinized use of the technology (Rogers 1995) where some authors have speculated about roles such as knowledge managers (Damsgaard and Scheepers 1999b).

Intranet technologies have a very large range of application environments. Fruitful areas for future research include the verification of the roles identified here in more cases, perhaps in different types of organizations (e.g., those in non-profit sectors) and also in different cultures. A further research avenue is to examine new and changed roles once the technology becomes routinized in the organization.

References

Ashby, W. R. *An Introduction to Cybernetics*. London: Chapman and Hall, Ltd., 1956.

Attewell, P. "Technology Diffusion and Organizational Learning: The Case of Business Computing," *Organization Science* (3:1), February 1992, pp. 1-19.

Bansler, J. P.; Havn, E.; Thommesen, J.; Damsgaard, J.; and Scheepers, R. "Corporate Intranet Implementation: Managing Emergent Technologies and Organizational Practices," in *Proceedings of the Seventh European Conference on Information Systems*, J. Pries-Heje, C. U. Ciborra, K. Kautz, J. Valor, E. Christiaanse, and D. Avison (eds.). Copenhagen, Denmark, June 23-25, 1999.

Beath, C. M. "Supporting the Information Technology Champion," *MIS Quarterly*, September 1991, pp. 355-371.

Beatty, C. A.; and Gordon, J. R. M. "Preaching the Gospel: the Evangelists of New Technology," *California Management Review* (33:3), 1991, pp. 73-94.

Beer, S. *Cybernetics and Management*, 2nd ed. London: The English University Press Ltd., 1959.

Benbasat, I.; Goldstein, D. K.; and Mead, M. "The Case Research Strategy in Studies of Information Systems," *MIS Quarterly* (11:3), September 1987, pp. 369-386.

Bernard, R. *The Corporate Intranet*. New York: Wiley & Sons, 1996.

Bhattacherjee, A. "Management of Emerging Technologies: Experiences and Lessons Learned at US West," *Information and Management* (33), 1998, pp. 263-272.

Clement, A. "Computing at Work: Empowering Action by 'low-level users'," *Communications of the ACM* (37:1), January 1994, pp. 53-105.

Damsgaard, J., and Scheepers, R. "Power, Influence and Intranet Implementation: A Safari of South African Organizations," *Information, Technology and People*, 1999a (forthcoming).

Damsgaard, J. and Scheepers, R. "A Stage Model of Intranet Technology Implementation and Management," in *Proceedings of the Seventh European Conference on Information Systems,*

J. Pries-Heje, C. U. Ciborra, K. Kautz, J. Valor, E. Christiaanse, and D. Avison (eds.). Copenhagen, Denmark, June 23-25, 1999b.

Dyson, P.; Coleman, P.; and Gilbert, L. *The ABCs of Intranets.* San Francisco: Sybex, Inc., 1997.

Eisenhardt, K. M. "Building Theories from Case Study Research," *Academy of Management Review* (14:4), 1989, pp. 532-550.

Ginzberg, M. J. "Key Recurrent Issues in the MIS Implementation Process," *MIS Quarterly,* June 1981, pp. 47-59.

Goles, T., and Hirschheim, R. (eds.). *Intranets: The Next IS Solution?* Houston, TX: Information Systems Research Center, College of Business Administration, University of Houston, 1997.

Hills, M. *Intranet Business Strategies.* New York: Wiley & Sons, 1997.

Howell, J. M., and Higgins, C. A. "Champions of Technological Innovation," *Administrative Science Quarterly* (35), 1990, pp. 317-341.

Humphrey, W. S. *Managing the Software Process.* Reading, MA: Addison-Wesley Publishing Company, 1989.

Jarvenpaa, S. L., and Ives, B. "Introducing Transformational Information Technologies: The Case of the World Wide Web Technology," *International Journal of Electronic Commerce* (1:1), 1996, pp. 95-126.

Keen, P. G. W. "Information Systems and Organizational Change," *Communications of the ACM* (24:1), 1981, pp. 24-33.

Klein, H. K., and Myers, M. D. "A Set of Principles for Conducting and Evaluating Interpretive Field Studies in Information Systems," *MIS Quarterly* (23:1), March 1999, pp. 67-92.

Kling, R., and Iacono, S. "The Control of Information Systems Developments after Implementation," *Communications of the ACM* (27:12), December 1984, pp. 1218-1226.

Lawless, M. W., and Price, L. L. "An Agency Perspective on New Technology Champions," *Organization Science* (3:3), 1992, pp. 342-355.

Lucas, H. C.; Ginzberg, M. J.; and Schultz, R. L. *Information Systems Implementation: Testing a Structural Model.* Norwood, NJ: Ablex Publishing Corporation, 1990.

Lyytinen, K.; Rose, G.; and Welke, R. "The Brave New World of Development in the Internetwork Computing Architecture (InterNCA) or How Distributed Computing Platforms Will Change Systems Development," *Information Systems Journal* (8), 1998, pp. 241-253.

Markus, M. L. "Power, Politics, and MIS Implementation," *Communications of the ACM* (26:6), June 1983, pp. 430-444.

Markus, M. L. "Toward a 'Critical Mass' Theory of Interactive Media: Universal Access, Interdependence and Diffusion," *Communication Research* (14:5), 1987, pp. 491-511.

Markus, M. L., and Benjamin, R. I. "Change Agentry: The Next IS Frontier," *MIS Quarterly,* December 1996, pp. 385-407.

Martinsons, M. G. "Cultivating the Champions for Strategic Information Systems," *Journal of Systems Management,* August 1993, pp. 31-34.

Oppliger, R. "Internet Security: Firewalls and Beyond," *Communications of the ACM* (40:5), May 1997, pp. 92-102.

Orlikowski, W. J.; Yates, J.; Okamura, K.; and Fujimoto, M. "Shaping Electronic Communication: The Metastructuring of Technology in the Context of Use," *Organization Science* (6:4), 1995, pp. 423-444.

Rogers, E. M. *Diffusion of Innovations,* 4th ed. New York: The Free Press, 1995.

Romm, C. T., and Wong, J. "The Dynamics of Establishing Organizational Web Sites: Some Puzzling Findings," *Australian Journal of Information Systems* (5:2), 1998, pp. 60-68.

Scheepers, R., and Damsgaard, J. "Using Internet Technology Within the Organization: A Structurational Analysis of Intranets," in *Proceedings of GROUP'97 Conference of Supporting Group Work*, S. C. Hayne and W. Prinz (eds.), Phoenix, AZ, Association for Computing Machinery, November 16-19, 1997, pp. 9-18.

Schön, D. A. "Champions for Radical New Inventions," *Harvard Business Review* (41:2), 1963, pp. 77-86.

Swanson, E. B. "Information Systems in Organization Theory: A Review," in *Critical Issues in Information Systems Research*, R. J. (Boland and R. A. Hirschheim (eds.). Chichester, England: Wiley & Sons, 1987, pp. 181-204.

Walsham, G. *Interpreting Information Systems in Organizations*. Chichester, England: Wiley & Sons, 1993.

Wynekoop, J. L., and Senn, J. A. "Case Implementation: The Importance of Multiple Perspectives," in *Proceedings of SIGCPR*, New York, Association for Computing Machinery, April 1992.

Yin, R. K. *Case Study Research: Design and Methods*. Newbury Park, CA: Sage Publications, 1989.

Zaltman, G.; Duncan, R.; and Holbek, J. *Innovations and Organizations*. New York: Wiley & Sons, 1973.

References

Rens Scheepers is a guest researcher in Information Systems at Aalborg University, Denmark where he is currently completing his Ph.D. His present research concerns the organizational dynamics associated with the implementation of intranet technology. He is on study leave from his position as Manager, Marketing and Business Systems at the CSIR in South Africa. He holds a B.Sc. (Hons.) in Computer Science and an MBA from the University of Pretoria. His URL is http://www.cs.auc/dk/~rens and his e-mail address is rens@cs.auc.dk.

13 A HERMENEUTIC INTERPRETATION OF THE EFFECT OF COMPUTERIZED BPR TOOLS ON REDESIGN EFFECTIVENESS IN TWO ORGANIZATIONS[1]

Suprateek Sarker
Washington State University
U.S.A.

Allen S. Lee
Virginia Commonwealth University
U.S.A.

Abstract

The business process reengineering (BPR) literature maintains that the use of computerized tools for BPR-related tasks such as process modeling, simulation, project management, and human resource analysis has a positive influence on the effectiveness of business process redesigns. Our hermeneutic study of text and text analogues surrounding BPR tool use in two organizations reveals that the use of computerized tools can have two opposing effects on redesign effectiveness. We find that, consistent with the existing BPR literature, BPR tools can indeed enhance redesign effectiveness by providing (1) a structure to the redesign process; (2) cognitive support to the redesigners; and (3) a mode for standardized representation of the redesigns. However, we also discover that the autonomization of electronically represented redesigns and the organizational members'

[1]The authors would like to thank John McKinney, Joseph Valacich, Dave Chatterjee, and Sundeep Sahay for their comments on earlier versions of this paper.

*subsequent focus on standardized, detailed, and objectified represen-
tations (rather than on socially shared understandings) of the
redesign, can lead to an alienation of the original redesigners from
the business processes that they envisioned. This alienation, coupled
with the redesigners' frustration arising from the frequent and
sometimes meaningless changes to the electronically objectified
redesigns mandated by other BPR stakeholders in the organization,
can contribute to inconsistencies in the redesign, thus resulting in a
negative influence of BPR tools on redesign effectiveness.*

*Our study (1) illustrates the use of the "hermeneutic circle" to
understand the role of computerized tools in business process
redesign; (2) argues that the role of computerized BPR tools can be
better understood by focusing on the sociotechnical interaction of the
redesigners with the computerized tools in an organizational context
rather than by studying the tools in isolation; and (3) indicates that
the effect of tools on redesign effectiveness depends on the relative
strengths of the two opposing effects.*

Keywords: Business process redesign, BPR tools, interpretive
methodology, hermeneutics, sociotechnical perspective.

1. Introduction

The academic and trade literature on BPR (business process reengineering) has
repeatedly noted that computerized BPR tools have a positive influence on the
effectiveness of the *product* of business process redesign,[2] as well as the *process* of
redesigning itself (Davenport 1993; Manganelli and Klein 1994; Klein 1998). The BPR
literature enumerates a number of alleged advantages of using these tools, but its claims
that these tools actually contribute to the effectiveness of redesign have not yet come
under empirical scrutiny. An objective of this study is to build on and advance the BPR
literature by establishing, based on empirical investigation, how BPR tools actually
contribute (or do not contribute) to effectiveness of business process redesign. To
achieve this, we[3] will (1) use the BPR literature itself to provide a starting point for an
understanding of BPR tools in business process redesign[4] and then (2) pursue an
interpretation of not only the *nature of influence* of computerized BPR tools on redesign

---

[2]Consistent with the BPR literature, we view business process reengineering as consisting of two
analytically separable phases: business process redesign (redesign), and the subsequent implementation of the
redesigned processes.

[3]For the sake of readability, we will use the first person plural throughout the paper. Whereas the first
author conducted all the fieldwork, the two authors collaborated in developing all the interpretations.

[4]There are other research literatures in the information systems field that can also provide starting points
for an empirical study of BPR. Examples are the literatures on information systems implementation and
structuration theory. However, because our investigation is making a point of contributing to the BPR
literature, it makes more sense to draw our theory from the BPR literature than from other literatures.

effectiveness, but also the *process* by which these tools influence redesign effectiveness. Specifically, we will study the use of computerized tools for process mapping/ flowcharting, project scheduling and other project-related documentation as part of the BPR initiatives in two corporations in a large city in the midwestern region of the United States.[5] We refer to the organizations as MANCO (a manufacturer of environmental products) and TELECO (a telecommunications service provider). Our methodology is hermeneutic, which is an interpretive approach for "reading" text and text-analogues (e.g., Lacity and Janson 1994; Lee 1994; Mueller-Vollmer 1994).

We intend a major contribution of our empirical investigation to be that it *refutes the past BPR literature's categorical presumption of a positive influence by BPR tools on the product and process of business process redesign.* To accomplish this, we use the BPR literature itself as our starting point for developing an understanding of BPR tool use at two organizations. Next, we show that what we actually observe regarding BPR tool use is at odds with what the BPR literature leads us to expect, whereupon we proceed in an iterative process (using the device of the hermeneutic circle) to identify deficient aspects in the BPR literature and to refine it into an interpretation that is able to account for how BPR tools were indeed used (or not used) in the redesign efforts at the two organizations.

In section 2 of this paper, we briefly review the literature on the role of computerized BPR tools. In section 3, we explain our interpretive methodology, some of the theoretical concepts that we use to interpret the use of computerized BPR tools, and our approach for evaluating redesign effectiveness. In section 4, we perform the interpretation using the hermeneutic circle and propose an improved way for understanding the role of BPR tools in redesign. Finally, in section 5, we discuss the contributions of this study and avenues for further research.

2. A Review of the Literature on the Role of BPR Tools

Articles and advertisements in BPR trade journals indicate that BPR tools will contribute to the effectiveness of the redesign in terms of cost, speed, ease of process-mapping, ease of redesigned process implementation and lowered project risk.[6] In the research-oriented BPR literature, Kettinger, Teng and Guha (1997) documented their findings from a study of 102 computerized BPR tools, concluding that "the tools survey indicates that an expanding suite of tools are being used to provide structure and information management

---

[5]The first author conducted the field work over a period of approximately 18 months in 1995 and 1996. The research involved both positivist and interpretive stages. To gather data, the first author followed the case study procedures of Yin (1994). After the positivist stage of the research (including hypothesis testing) was completed, the case study's empirical material (including the interview material) then became the "text" for a hermeneutic interpretation. For additional details, see Sarker (1997).

[6]For example, see *Enterprise Reengineering* (2:5), August 1995.

capability[7] in conducting BPR techniques and possess the potential to accelerate BPR projects" (p. 63).

Consistent with this point of view, Carr and Johannson (1995) explain and illustrate the importance of using such tools, and propose the following "prospective best practice" for BPR initiatives: "Take advantage of modeling and simulation tools" (p. 150). Further highlighting the importance of computerized tools, Davenport (1993, p. 216) points out three "paramount" dangers associated with the failure to pursue opportunities provided by advanced technological tools: first, failures to employ these tools can reduce the pace at which the redesign will progress and this, in turn, can reduce the chance of the initiative's succeeding; second, these tools are likely to improve the quality of the product of redesign; and third, it could indicate that managers are not aware of technological opportunities, consequently undermining the importance of the initiative. Finally, Klein (1998, p. 245) also recognizes the importance of BPR tools, stating:

> By using tools, the BPR practitioner expects to improve productivity, finish projects faster, produce higher quality results and eliminate tedious housekeeping work in order to concentrate on value-added work. To produce these benefits, BPR tools should be useable by businesspeople (managers and professionals), not technicians.

An important aspect of useability is "learnability," which Manganelli and Klein (1994) further highlight. They caution that while several benefits (such as "improved productivity," "faster projects," "higher quality levels," and "elimination of tedious work") can be expected from reengineering tools, "these benefits come only after first learning the tool" (p. 214).

To summarize, *the BPR literature is unequivocal in its position that computerized BPR tools have a positive influence on redesign effectiveness*, especially if the tools are easy to learn and use, where the positive influence follows from the fact that the tools provide (1) a structure for the BPR initiative; (2) cognitive support to the process designers (through the tools' information management capability); (3) a means of documentation (process diagrams, E-R diagrams, Gantt charts, etc.) for enabling communication among different stakeholders of the reengineering initiative; and (4) a means of simulation and playing "what-if" in order to reduce the risks associated with a BPR initiative.

3. Data Collection, Research Methodology and Theoretical Concepts

3.1 Data Collection and Context

Our empirical material ("data") consists of formal interviews *on various aspects of BPR* with over 22 BPR stakeholders in the two organizations (MANCO and TELECO) that

---

[7]Kettinger, Teng and Guha view BPR tools with repositories and data indexing features to facilitate "collective knowledge sharing" as having information management capability.

we tape-recorded and transcribed. In addition, our empirical material also consists of informal conversations with stakeholders and observations of organizational members at work in their "natural settings." We provide the case summaries of TELECO and MANCO in this paper's appendix. We examined the empirical material collected from the two organizations through the interpretivist lens of *hermeneutics*.

3.2 Methodology

Hermeneutics originated as an approach for interpreting ancient religious texts that were alien to their contemporary readers, this alienation resulting from the historical and cultural distance between the readers and the authors of the text. Since then, contemporary social scientists have appropriated and suitably modified hermeneutics for the purposes of understanding not only "everyday" written text, but even speech acts and overall human behavior as *text analogues* (Davis et al. 1992; Ricoeur 1981). In this study, we treat utterances in interviews or informal conversations, and also actions of the BPR stakeholders in the two organizations, as text analogues. These texts were not static but continued to change in meaning with the occurrence of additional events in the organizations over time, and with the evolution in our own interpretation of the events through the use of the "hermeneutic circle."

The hermeneutic circle is a device that allows the reader to comprehend the parts of a "text" in terms of the whole, and the whole in terms of the parts (Davis et al. 1992; Geertz 1983). Perhaps, Thomas Kuhn best illustrates the notion of the hermeneutic circle when he states (see Lee 1991, p. 348):

> When reading the works of an important thinker, look first for the apparent absurdities in the text and ask yourself how a sensible person could have written them. When you find an answer...when those passages make sense, then you may find that more central passages, ones you previously thought you understood, have changed their meaning.

There are many schools of thought within the tradition of hermeneutics (Mueller-Vollmer 1994; Palmer 1979; Smith 1993), where the different schools of thought result from the polarizations of scholarly communities based on ontological and epistemological differences. While we recognize that there are merits to each school, we adopt an approach that is closely related to the school of *validation hermeneutics* developed primarily by Hirsch (1967) drawing on Dilthey's conception of the hermeneutic circle as a *methodological* device (Burrell and Morgan 1979; Pressler and Dasilva 1996). Smith (1993, p. 191) has characterized this approach as the most objectivist of the hermeneutic schools in that it assumes that "inquiry is pointless and the concept of knowledge makes no sense in the absence of an independently existing entity to inquire about or have knowledge of."[8] Smith (pp. 191-192) describes the essence of validation hermeneutics, drawing on Hirsch's work:

> An inquirer begins with a hypothesis (or hypotheses) about meaning and then searches for evidence that will call the hypothesis into doubt. If such a falsification evidence is uncovered, the inquirer must revise the interpretation. Throughout the process of constantly testing one's interpretation of meaning, "the direction is still toward increased probability of truth, since the very instability imposed by unfavorable evidence reduces confidence in a previously accepted hypothesis and to that extent reduces the probability of error" (Hirsch, 1967, pp. 151-152)....The process of interpretation cannot be reduced to a rule-bounded or mechanical process. However, this absence of rules does not mean that "anything goes" because the attempt to interpret an author's meaning is constrained by constant testing, criticism, and so on in the name of the search for truth.

Clarifying the role of validation in hermeneutic interpretation, Hirsch (1967, p. 170) further adds:

> The exigencies of validation are not to be confused with the exigencies of understanding....Every interpretation begins and ends in a guess, and no one has ever devised a method for making intelligent guesses. The systematic side of interpretation begins where the process of understanding ends. Understanding achieves a construction of meaning: the job of validation is to evaluate the disparate constructions which understanding has brought forward. Validation is therefore the fundamental task of interpretation as a discipline.

He cautions, however, that any consensus regarding an interpretation achieved through validation is necessarily temporary, and changes as new facts and guesses appear. Consistent with this point-of-view, Davis et al. (1992, p. 304), offer the following guidelines for assessing the "quality of an interpretation":

> A "good" interpretation resolves any apparent anomaly or irrationality. A good interpretation, however, need not be (and, in fact, cannot be) final and conclusive because, at least in principle, improvements in the interpretation will always be pursuable.

Thus, in our study, while we recognize that further passes through the hermeneutic circle can result in an improved interpretation, we may discontinue the circular motions around the organizational "text" when we are satisfied by our latest interpretation and are not left confronting any glaring anomalies or "apparent absurdities."

------

[8]Note that an objectivist ontology does not necessarily entail an objectivist methodology. There can be multiple interpretations of the same independently existing entity in social research, just as there can be competing theories of the same independently existing entity in natural-science research.

3.3 Theoretical Concepts Aiding the Interpretation

While performing the interpretation, in addition to the above guidelines, we also found the hermeneutic concepts of *distantiation, autonomization, and appropriation* (Boland 1991; Lee 1994; Ricoeur 1981), and the notion of *mutual understanding* (Churchman and Schainblatt 1965) useful in creating a processual understanding of the use of computerized tools for redesigning business processes. We offer, as background, basic definitions of these concepts, which our subsequent investigation uses. *Distantiation,* in the context of the redesign efforts at MANCO and TELECO, refers to the separation or the "disconnect" that occurs between a particular "text" (i.e., electronic representation of redesigned processes) and its authors (i.e., the redesigners of the business processes). *Autonomization* refers to the life that the electronic (textual/graphical) representation of redesign takes on, independently of its original form and even of the intentions of its authors. *Appropriation* refers to the process by which a party in one socially-constructed world comes to understand the "text" (i.e., redesigns) created in yet another socially-constructed world. *Mutual understanding* involves a dialectical and recursive process (rather than uni-directional processes of communication or persuasion) through which the redesigners and the managers come to understand each other's interpretation of the redesigns. In addition to these concepts, we wholeheartedly adopted the hermeneutic principle that any apparently irrational behavior of people is not a sign that people are irrational, but that they are responding in rational ways to the (likely irrational) circumstances of their context, which would then call for interpretation by us, the researchers (Davis et al. 1994; Lee 1991).

3.4 Evaluation of Redesign Effectiveness

Before proceeding with the hermeneutic interpretation, we briefly clarify our approach for evaluating redesign effectiveness in this study. Evaluation of the effectiveness of redesign in BPR is a complex activity, much like the evaluation of information systems implementation success, and no universally accepted criteria exist for such evaluations (Boudreau and Robey, 1996). While several criteria for evaluation of BPR have been discussed in the literature (e.g., Boudreau and Robey 1996; Sethi and King 1998), we believe, drawing on Lyytinen and Hirschheim's notion of "expectation failure" (1987, p. 264), that an assessment of redesign effectiveness requires the recognition of the existence of multiple stakeholders of the redesign initiative, having different values, levels of power and interests, and hence, different expectations; thus, a thorough examination of the evaluations of the various stakeholders of the initiative is necessary. For the purpose of this study, we consider redesign to be effective *if different stakeholders state or indicate through actions* that such was the case. The fact that we are tying the notion of redesign effectiveness to subjective meaning (i.e., the meanings and perceptions held by the human subjects whom we observe in our study) therefore makes an *interpretive* research approach appropriate. Furthermore, because we are interpreting manifestations of our human subjects' meanings in the forms of text (their utterances) and text analogues (their actions), *hermeneutics* is a suitable interpretive approach for this study.

4. The Interpretation

Our interpretation using the hermeneutic circle involved several "circular passes" around the text, with each such iteration ending with a different understanding and also a different puzzle, thus bringing a different set of "data" to our (i.e., the researchers') focus. Each pass through the hermeneutic circle involved four broad steps, which we derive from the work of Davis et al.: first, the identification of breakdowns in our understanding as researchers, resulting from a contradiction between what we already understand (sometimes this is called the "pre-understanding") and what we actually observe; second, the examination of new data relevant to the breakdown being investigated and/or the reexamination (in a new light) of data examined in a previous iteration; third, the surfacing of questionable assumptions (we had made earlier) that contributed to the breakdown in our understanding; and fourth, our revision of the existing interpretation to resolve the breakdown.

4.1 Passes Through the Hermeneutic Circle

The first iteration:

Our objective in the first pass of the hermeneutic circle was to verify our interpretation of the BPR literature regarding computerized tools. Our pre-understanding, based on the BPR literature, was that computerized BPR tools (especially those for flowcharting/ process-mapping and project management) enhance redesign effectiveness by providing (1) a necessary *structure* to the complex redesign process involving multiple redesigners over an extended period of time; (2) *cognitive support* to the redesigners who are overwhelmed by the amount of information and the linkages between them; and (3) a *standardized/shared notation* for representing business processes and other related information.
     In the case of MANCO, we learned that the MIS manager had acquired an *easy-to-learn user-friendly computer-aided flowcharting tool* specifically for use during the information-technology-enabled redesign phase of the reengineering initiative, and that this tool was used at MANCO in the early stages of this phase. However, the use of this tool was discontinued soon after the redesign team[9] started to meet for the purpose of envisioning how MANCO's information-technology-enabled business processes should work. The redesign team at MANCO accomplished the business process redesign through a process lasting several months in which the team iteratively brainstormed, discussed, and agreed upon different aspects of future business processes and the organization around those processes. Because the redesign team discontinued its use of the flowcharting tools, the evolving redesign primarily existed in the minds of the redesigners in the form of a shared body of knowledge that was not represented as

---

[9] We use the terms "redesign team" and "redesigners" to refer to those organizational members who were formally involved in the rethinking and envisioning of the new business processes and the related organizational aspects. This terminology is consistent with the BPR literature, which breaks down BPR into a redesign stage (in which business processes are redesigned) and a subsequent implementation stage (in which new designs are put into effect).

written or computer-represented text or diagrams. On some occasions, especially for clarification purposes, the redesigners spontaneously hand-drew flowcharts in redesign sessions,[10] but *at no point, was there any attempt to create computer-drawn process diagrams representing the team's then current vision of any business process, even though MANCO had a flowcharting software readily available.* Toward the end of the redesign effort, we asked the MIS Manager why the process redesigns had not been represented using the flowcharting package. She said:[11, 12]

> *I had tried to do that...it just worked out to be an exercise for me... basically. If you look in my book that I put together before the project started, I had... two chapters... "business as it is" and "business as it will be," and the "will be" is still blank. The vision that we have right now is kind of a high level and it hasn't really come to fruition yet...we will write (draw) it after we do it.*

She was convinced that the use of computerized flowcharting tools would not contribute to a more effective redesign, especially in light of the iterative approach to process redesign that the reengineering team had adopted. According to the MIS manager, an important advantage of not using the flowcharting tool was that the design could then remain very flexible and could also be continuously challenged and modified by the team members, who continually encountered different concerns as they learned more about the processes being redesigned and about the process-enabling software options. When asked if she would have used BPR tools in a larger company, the MIS manager indicated that she probably would have to, although (interestingly) not because such tools would *inherently* enhance the effectiveness of the process redesigns, but because they could help generate the "professional documents" and contribute to the legitimacy of the redesigners in larger organizations:

> *In a larger company you have to justify things a lot more....And you have to get sign-offs and go through the levels of approval and all this stuff...but here, it's not like that.*

All MANCO redesigners whom we interviewed expressed the view that the use of computerized tools would not have enhanced redesign effectiveness. Our own observation during the redesign sessions also supported the team-members' shared view that the absence of computerized graphical tools helped the team to operate flexibly without getting bogged down on details and diagraming conventions.

---

[10]Even the hand-drawn flowcharts were not saved for future use; they were drawn for the purpose of explanation and clarification only.

[11]When quoting organizational members, we place the text in italicized font. Furthermore, because the same text can come to have different meanings in different passes through the hermeneutical circle (please also refer to the quotation of Kuhn, above), we will additionally place some portions of the text in bolded font to help emphasize the particular meaning we are forming at the current stage in our interpretation.

[12]Space limitations preclude us from presenting a full complement of the organizational members' remarks and other empirical material. It is available in Sarker (1997).

The TELECO reengineering team members, on the other hand, reported extensive use of software for flowcharting/process-mapping and also for project management (although not for simulation). In general, the TELECO redesigners had a positive disposition toward such tools and appeared to believe that the tools had indeed contributed to a better redesign of the business processes. A TELECO redesigner said:

> *[W]e used Microsoft Project, a flowcharting software...Visio flow-charting software, WordPerfect documents...and PowerPoint....I* **would say Visio and Project helped us the most during the redesign phase.** *We used Visio to create all the process flowcharts....it was just fantastic...and Project...we really stretched its capabilities and used it to integrate plans across all the people involved....I would say that* **the design would not have been as effective without the use of the tools.**

Based on the interviews with the redesigners, we concluded that computerized BPR tools had had a positive influence on the effectiveness of the business process redesign at TELECO.

In summary, while we found TELECO's experience to be consistent with the literature on BPR tools, MANCO's experience contradicted our understanding of this literature. In particular, at MANCO, we observed that (1) the MIS manager discontinued the use of a user-friendly and easy-to-learn flowcharting tool that had been acquired specifically to facilitate the redesign and (2) the redesign team members stated that the tools would not have made the redesign more effective (and, in fact, could have contributed negatively to effectiveness by reducing flexibility). The only interpretation that we could offer at this point was that MANCO's use of BPR tools was different in some unique way due to which the organization had experienced a negative effect of the tools. However, even we ourselves were not quite satisfied with this particular interpretation.

The second iteration:

In an effort to resolve the apparent breakdown, we returned to the organizations, asking the redesigners questions that we hoped would prod self-reflection among them. We hoped that some additional organizational "text" would reveal itself and help us make sense of the breakdown in our understanding. Finally, a redesigner at TELECO provided a way for us to resolve the breakdown. In the course of an interview, he burst out unexpectedly:

> *The problem is, if you have a tool, you become a slave to that tool....* **we did more damn presentations, to try and get a buy into what we were doing, that we spent too much time....The business of produc-ing and documenting was very cumbersome...**we refined the hell out of this thing...and tool-smithed it so many times, it was ridiculous!*

This outburst of the TELECO redesign team member immediately led us to question our starting assumptions. We had assumed that process losses during redesign initiatives (those that are not supported by computerized tools) occur primarily due to cognitive limitations of team members, due to confusion in the process of redesign, or due to the lack of a standard notation for representing existing and envisioned business processes. However, in the case of TELECO, while the tools appeared to have contributed in all three areas mentioned above, we were no longer convinced (given the outburst of the redesigner) that the tools had positively contributed to the redesign. Consequently, our conclusion at the end of this iteration was that BPR tools have a negative effect on redesign effectiveness![13]

The third iteration:

Yet, the interpretation that computerized BPR tools have a negative effect on the redesign seemed extremely counter-intuitive. In light of so much "evidence" in the trade and academic journals regarding the positive experiences of BPR tool users, this new interpretation did not seem to ring true to us, presenting us with an anomaly that had to be explored further. We recalled that one of MANCO's redesigners, who had provided us with many insights through his critical thinking, had mentioned his uneasiness with the lack of structure in the design process:[14]

> It [the flowcharting tool] would have provided us with **some guid-
> ance.**

This made us question yet another *assumption* that we had unwittingly used in this interpretation: computerized graphical tools affect redesigns in *only one direction*. Was it not conceivable that BPR tools can have both a positive influence and a negative influence on redesign effectiveness?

Fourth iteration:

At the starting point of the fourth iteration, we found ourselves confronting two breakdowns in our understanding regarding the use of BPR tools at MANCO and TELECO. The first breakdown was related to our understanding that the BPR tools do have (in certain circumstances) a positive influence on redesign effectiveness by providing structure, standard notation and cognitive support. Because structure, standard notation and cognitive support become more relevant with increasing size, it seemed sensible for us to assume that a larger reengineering team within a larger organization would experience the benefits from using the tools. Yet, as the redesign team member's outburst revealed, there was a significant level of dissatisfaction (and by implication,

---

[13]After all, MANCO also appeared to have experienced a somewhat negative influence of BPR tools.

[14]This particular redesigner too had said that overall the use of flowcharting tools would not have made the redesign at MANCO more effective.

negative influence on redesign effectiveness) in TELECO (which was the larger of the two organizations with a considerably larger reengineering team) regarding the use of computerized tools.

The second breakdown that we were experiencing was that we did not understand why MANCO had discontinued the use of the BPR tools that it had acquired even though a respected member of the redesign team had expressed the need for a computer-based redesign support tool.

In attempting to resolve these breakdowns, we re-examined what a TELECO reengineering team-member had said when reflecting on his experience with the tools:[15]

> The problem is, if you have a tool, **you become a slave to that tool**....
> we did more **damn presentations** to try and get a buy into what we
> were doing, that we spent too much time...I mean producing those
> things. The business of producing and documenting was very
> cumbersome...**we refined the hell out of this thing...and toolsmithed
> it so many times, it was ridiculous**....they're only as good as how
> people follow them, because if there is no real dedication to plans...
> all the tools in the world won't.

This expression of frustration indicated to us that the computerized tools for representing redesigns could cause an alienation (or distantiation, in the hermeneutic sense) of the designs from its original formulators. We could also see that much of the unhappiness experienced by the redesign team member was due to the fact that he had been forced to make "meaningless" changes.

In attempting to resolve our second breakdown, two quotations that we had previously examined came alive in a different way, triggering an interpretation that could resolve the contradiction:

> MANCO redesigner: *It would have provided us with **some guid-
> ance...initially**.*

> MANCO MIS manager: *In a larger company, you have to **justify
> things a lot more**....and you **have to get sign-offs and go through the
> levels of approval and all this stuff**.*

On examining these "data," we were even more convinced that BPR tools do not influence redesign effectiveness in one direction only. Also, we realized that, in addition to providing a structure, some cognitive support, and a mode for standardized representation, computer-based BPR tools serve another important social function in the context of BPR: that of helping to justify a redesign through the layers of bureaucracy by creating a formal or "professional" appearance of the redesign.

---

[15]Our return to a text, which we have already examined, is deliberate. In hermeneutic interpretation, "[w]hen you find an answer...then you may find that more central passages, ones you previously thought you understood, have changed their meaning" (Kuhn, quoted above). Such a return does not "privilege" a given text (or the person who authored it), but assures that it receives due consideration in fitting into a coherent, overall interpretation. This stands in contrast to an approach based on random sampling, in which each datum would "count" equally and only once.

The revised understanding emerging toward the end of the fourth pass was that the BPR tools can actually be helpful in the early stages of the redesign process in providing a structure and buffer against the redesigners' cognitive limitation of comprehending and retaining unfamiliar process related designs immediately. The tools could also help in creating "professional-looking" redesigns that could be presented to management for approval. In the later stages, this very structure and the ability to create professional-looking redesigns could act as a constraint to creative thinking, and could instead encourage a bureaucratic mind-set with an obsession on maintaining consistency in the diagrams and other unimportant details. Our new interpretation of the situation better accounted for the dissatisfaction of the TELECO redesigner as well as MANCO's decision to discontinue the use of computer-based BPR tools.

The fifth iteration:

Our interpretation, up to this point, seemed to suggest that BPR tools should be used only in the early stages of the redesign or to satisfy bureaucratic requirements; yet, many organizations (including ones that cannot be characterized as "bureaucratic") have been reported to use the tools effectively throughout the life cycle of the initiatives. To address this issue, we revisited some "data" that we had used for surfacing an interpretation for a previous iteration.

> The MANCO MIS manager: *In a larger company, you have to justify things a lot more....and you have to get sign-offs and go through the levels of approval and all this stuff...but here, it is not like that...* at least not now... it used to be.

> A TELECO redesign team member: *The problem is, if you have a tool, you become a slave to that tool....we did more damn presentations to try and get a buy into what we were doing, that we spent too much time....The business of producing and documenting was very cumbersome...we refined the hell out of this thing...and toolsmithed it so many times, it was ridiculous....they're only as good as how people follow them, because if there is no real dedication to plans... all the tools in the world won't help.*

We noticed that specific words in the text newly stood out as we searched for a way to resolve the anomaly confronting us. For example, we found the MIS manager's words "In a large company, you have to justify things lot more...but *here, it is not like that...* at least for now...it used to be" and the TELECO redesigner's words "you *become* a slave to that tool....we refined the hell out of this thing...it was ridiculous....If there is no dedication to plans...all the tools in the world won't help" to be prominent in our view. The interpretation that we now crafted was that decisions to use BPR tools, and the effects of BPR tools, are dependent on *the context of their use.* Tools can provide much needed structure and cognitive support throughout the life of the project; however, care must be taken so that an alienation of the authors from the design does not occur. Too much emphasis on the tools may result in the redesign (represented in electronic form) being taken-for-granted as an "objective" product that is distantiated from the redesigners and autonomized. The autonomized design could then subsequently end up in the hands of different stakeholders, in different socially constructed worlds, who initiate corrections and modifications even when they have no awareness of the original context of the redesign. We also came to appreciate that a standardized representation of the redesign does not guarantee a uniform interpretation/appropriation across different stakeholder groups (owing to the different socially constructed worlds to which different groups belong, and the different frames that they use to interpret a given representation [Orlikowski and Gash 1994]), and also that a shared social context and mutual understanding among different designers and management is key to a positive influence of computerized BPR tools on redesign effectiveness.

Another point worthy of emphasis was that we were beginning to see the *redesigners' interaction with the tools in a particular context* (and not the tools themselves) as influencing the redesign in a positive or a negative manner. Interestingly, this indicated a shift in the underlying theoretical orientation of the understanding that was evolving through the interpretation, from a technological-imperative orientation to a sociotechnical orientation (Markus and Robey 1988).

At this point, it appeared to us that we had a satisfactory understanding; we had validated it to the extent our text allowed us and there were no longer any anomalies or breakdowns confronting us. Thus, we decided not to undertake further interpretive iterations. The processes of redesign at TELECO and MANCO, as framed in our hermeneutic interpretation, are summarized in Figure 1.

4.2 Putting the Pieces Together: A Recapitulation of Insights Gained Through Interpretation

Our interpretation revealed that computerized BPR tools do provide a structure to organize redesign activity and a common language to aid redesign team members in communicating and sharing their emerging visions, both among themselves and with other stakeholders such as the members of top management. The tools provide a buffer for team members against their cognitive limitations of comprehending information about unfamiliar business processes (whether existing or envisioned). However, once the team members become sufficiently familiar with the existing processes and visions, the "structure" of the existing designs, whether in an electronic medium or on paper, can act as a constraint to creative thinking about the processes. As a result of this "structure," redesign team members no longer engage in creative thinking in an *ad hoc* fashion, but instead tend to adopt a bureaucratic mind-set that leads them to focus their attention on maintaining consistency in the diagrams and arguing about unimportant details. Also, the process diagrams, once formalized on paper or in electronic media, become distantiated from their authors and thereafter become autonomized. Consequently, different redesign team members, owing to different perspectives associated with their different social contexts, can come to understand the same redesign differently, and can accordingly make changes to these formalized redesigns/diagrams, leading to an alienation of the redesigns from their original authors. This loss of control and personal ownership, in turn, can lead to the original authors losing interest and commitment to the redesign.

Also, when this redesign is presented to top management, who typically exist in a socially constructed world different from that of the redesign team members, they (the

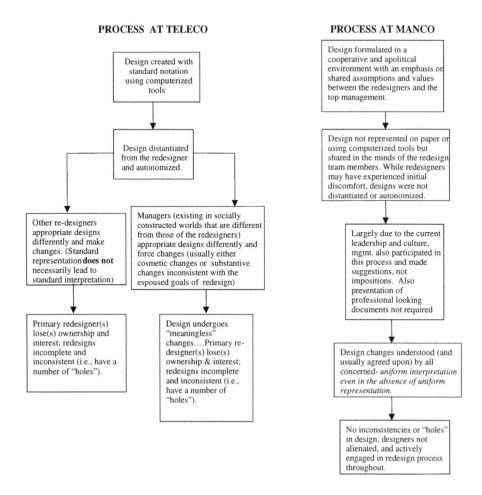

Figure 1. The Processes of Redesign

members of top management) can then appropriate the redesign within their own context, thus obtaining yet another different understanding of the redesign as compared to the understanding of the redesign authors. This is evident in TELECO where the redesigners were keen to show how cross-functional coordination would be enabled by their envisioned process while management was instead interested to see how many head-counts had been reduced, or how many process owner positions (for themselves) had resulted from the redesign effort. Having more power than the redesigners in determining the final product of redesign, management imposed changes and refinements that were, often times, not important to the business process redesign and, thus, meaningless to the redesigners; yet, the redesign team members had no choice but to comply helplessly with those directives resulting in the kind of bottled-up frustration captured in their own usage of words such as "slave," "damn presentations," "refined the hell out of," "ridiculous," and "all the tools in the world won't help." It is clear that making changes to the redesign had become a detached task for the TELECO redesigner—a task that the redesigner then had to accomplish in order to satisfy someone else's wishes, as opposed to making changes for "improving our design." Such distantiation of the redesign document from its authors (the redesigners), and the alienation experienced during the process of redesign, both serve to account for our observations of glaring shortcomings in the redesign called "holes" (incompleteness and inconsistency in process redesigns) that became apparent during the later implementation of the redesigned processes at TELECO.

In contrast, while MANCO redesigners may have experienced some initial difficulty in comprehending the organizational business processes and getting accustomed to the redesign project, subsequently their personal involvement and their relatively apolitical environment guaranteed that the redesign was a part of their respective and also shared stocks of knowledge. By not having the redesign on paper or in electronic media, MANCO stood the risk of not having any *standardized representation* at all, although having a standardized representation does not, by any means, guarantee a standardized interpretation. In fact, the taken-for-granted objectivity of electronically represented process maps could lead to less effort in developing a shared understanding of the redesign among team members, thereby causing serious confusion in the team's redesign efforts.

MANCO's organizational context, instead, encouraged a *standardized interpretation* of a mentally-shared conceptual representation of the processes among the redesign team-members. In addition, because the senior VP and other managers were closely participating in the project, they too were part of the same socially constructed world of the redesigners, and thus shared the same interpretation of the emerging redesigns. Whenever redesign changes had to be conveyed to the management, it was done personally, through a process of mutual understanding (i.e., the interpretation with immediate clarifications from both sides) rather than through the process of "half-duplex" communication that occurred in TELECO, involving a uni-directional presentation from the redesign team members followed by uni-directional change directives from the management.

The graphical representation (Figure 2) shows the *two opposing impacts of the computerized tool use on redesign effectiveness*: a positive impact arising from the structure, the cognitive support, and the standards for process representation provided by the tools; and a negative impact arising from narrow interpretations (by organizational members) of standardized representations of processes, the depersonalization of the reengineering vision, and the alienation of the vision from the redesigners. We emphasize that, in Figure 2, we do not intend the curves to serve as *precise* depictions of the forces that influence redesign positively and negatively. Rather, we intend them to show that the direction of the overall effect of BPR tools on the redesign effectiveness (depicted by curve C) depends on the relative strengths of the two opposing forces (depicted by curves A and B respectively).

5. Conclusion

We believe that our study makes two important contributions. First, methodologically, it illustrates how the hermeneutic circle can aid in the development of a "good" interpretive understanding (Davis et al. 1992) of an IT-related social phenomenon. Second, through a hermeneutic interpretation of organizational text surrounding

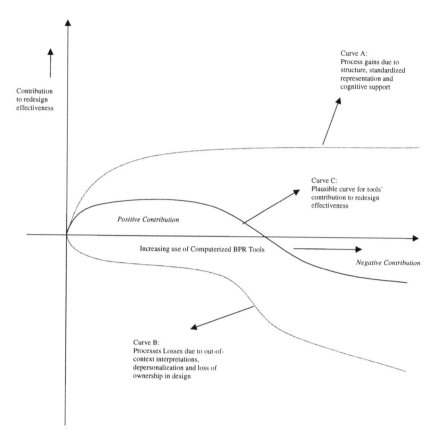

**Figure 2. A Preliminary Understanding of BPR Tool Use
Developed from the Interpretive Study**

computerized BPR tool use (or lack of use) in two organizations, we have learned that BPR tools need to be studied from a sociotechnical perspective, and that the overall direction of influence of a BPR tool on redesign effectiveness depends on whether or not *the context and degree* of the tool's use result in the positive impacts (depicted in Curve A, Figure 2) dominating the negative impacts (depicted in Curve B, Figure 2), or *vice versa*. We have also come to an understanding of the process by which the negative impact of computerized BPR tools could have occurred or could have been avoided (Figure 1).

While our hermeneutic study provides an alternative understanding of how BPR tools influence redesign effectiveness, more research needs to be done in this area. We see future research on this topic taking two inter-related forms (Lee 1991): for researchers seeking an even deeper understanding of the processes that enable and constrain redesign, a fruitful way to pursue research would be to conduct further interpretive examinations of BPR tools in organizational contexts using ethnographic or hermeneutic approaches; for researchers seeking findings that fit the positivist variants of the concepts of generalizability and generality, we recommend the use of deductive positivist studies that attempt to "operationalize" insights gained from the intensive interpretive examination of BPR tools, as in this study. These suggestions pertain to theory refinement and development, which would address the problem of the paucity of theory in the existing BPR literature. Efforts to build theory in BPR can, of course, also benefit from targeting other established areas of information systems research. For instance, now that our interpretation has established the significance that can characterize the interactive nature of the relationship between computerized BPR tools and their social context, it would appear appropriate and fruitful for future research on the effectiveness of BPR tool use to avail itself of insights from the literatures of sociotechnical systems theory and adaptive structuration theory.

The desirability of refining our hermeneutic interpretation with research literatures in addition to that of BPR, the necessity of continuing the process of interpretation (whether with observations at the same firms, MANCO and TELECO, or at additional sites), and the option of even beginning the interpretation process anew (for instance, embarking with a pre-understanding not from the BPR literature, but from the IS implementation literature) all serve to highlight the fact that alternative, and competing, interpretations are always possible and even welcome. Improvements in a hermeneutic interpretation are always possible, but the reader (the interpreter) may rest once he or she is satisfied that the latest reading (interpretation) has resolved the remaining anomalies or "apparent absurdities" (Kuhn, quote above) in the text or text analogue.

References

Burrell, G., and Morgan G. *Sociological Paradigms and Organizational Analysis*. London: Hienemann, 1979.

Boland, R. J. "Information System Use As A Hermeneutic Process," in *Information Systems Research: Contemporary Approaches and Emergent Traditions*, H. E. Nissen, H. K. Klein, and R. Hirschheim (eds.). Amsterdam: Elsevier-North Holland, 1991, pp. 439-457.

Boudreau, M., and Robey, D. "Coping with Contradictions in Business Process Re-engineering," *Information Technology and People* (9:4), 1996, pp. 40-57.

Carr, D. K., and Johansson, H. J. *Best Practices in Reengineering: What Works and What Doesn't in the Reengineering Process*. New York: McGraw-Hill, 1995.

Churchman, C. W., and Schainblatt, A. H. "The Researcher and The Manager: A Dialectic Of Implementation," *Management Science* (11:4), February, 1965.

Davenport, T. H. *Process Innovation: Reengineering Work Through Information Technology*. Boston: Harvard Business School Press, 1993.

Davis, G. B.; Lee, A. S.; Nickles, K. R.; Chatterjee, S.; Hartung, R.; and Wu, Y. "Diagnosis of an Information System Failure: A Framework and Interpretive Process," *Information and Management* (23:5), 1992, pp.293-318.

Geertz, C. "From the Native's Point of View: On the Nature of Anthropological Understanding," in *Local Knowledge*, C. Geertz (ed.). New York: Basic Books, 1983, pp. 55-70.

Hirsch, E. D. *Validity in Interpretation*. New Haven, CT: Yale University Press, 1967.

Lacity, M. C., and Janson, M. A. "Understanding Qualitative Data: A Framework of Text Analysis Methods," *Journal of MIS*, Fall 1994, pp. 137-155.

Lee, A. S. "Electronic Mail as a Medium for Rich Communication: An Empirical Investigation Using Hermeneutic Interpretation," *MIS Quarterly* (18), June 1994, pp. 143-157.

Lee, A. S. "Integrating Positivist and Interpretivist Approaches to Organizational Research," *Organization Science* (2:4), 1991, pp. 342-365.

Kettinger, W. J.; Teng, J. T. C.; and Guha, S. "Business Process Change: A Study of Methodologies, Techniques and Tools," *MIS Quarterly*, March 1997, pp. 55-80.

Klein, M. M. "Reengineering Methodologies and Tools: A Prescription for Enhancing Success," in *Organizational Transformation through Business Process Reengineering,* V. Sethi and W. R. King (eds.). Upper Saddle River, NJ: Prentice Hall, 1998, pp. 243-251.

Lyytinen, K., and Hirschheim, R. "Information Systems Failures: A Survey and Classification of the Empirical Literature," *Oxford Surveys in IT* (4), 1987, pp. 257-309.

Manganelli, R. L., and Klein, M. M. *The Reengineering Handbook: A Step-By-Step Guide to Business Transformation.* New York: AMACOM, 1994.

Markus, M. L., and Robey, D. "Information Technology and Organizational Change: Causal Structure in Theory and Research," *Management Science* (34:5), May 1988, pp. 583-598.

Mueller-Vollmer, K. *The Hermeneutics Reader.* New York: Continuum, 1994.

Orlikowski, W. J., and Gash, D. "Technological Frames: Making Sense of Information Technology in Organizations," *ACM Transactions on Information Systems* (12:2), April 1994, pp. 174-207.

Palmer, R. E. *Hermeneutics: Interpretation Theory in Schleiermacher, Dilthey, Heidegger, and Gadamer.* Evanston, IL: Northwestern University Press, 1979.

Pressler, C. A., and Dasilva, F. B. *Sociology and Interpretation.* New York: State University of New York Press, 1996.

Ricoeur, P. *Hermeneutics and the Human Sciences,* J. B. Thompson (ed.). New York: Cambridge University Press, 1981.

Sarker, S. *The Role of Information Technology and Social Enablers in Business Process Reengineering: An Empirical Investigation Integrating Positivist and Interpretivist Approaches,* Unpublished Doctoral Dissertation, Department of Accounting and Information Systems, University of Cincinnati, 1997.

Sethi, V., and King, W. R. (eds.). *Organizational Transformation Through Business Process Reengineering.* Upper Saddle River, NJ: Prentice Hall, 1998.

Smith, J. K. "Hermeneutics and Qualitative Inquiry," in *Theory and Concepts in Qualitative Research: Perspectives from the Field,* D. J. Flinders and G. E. Mills (eds.). New York: Teachers College Press, 1993.

Yin, R. K. *Case Study Research: Design and Methods.* Beverley Hills, CA: Sage, 1994.

About the Authors

Suprateek Sarker is an Assistant Professor of Information Systems at the Washington State University. He received his Bachelor of Computer Science and Engineering from Jadavpur University (India), MBA from Baylor University, M.S. in CIS from Arizona State University, and Ph.D. in Information Systems from the University of Cincinnati. Currently, he teaches undergraduate and graduate courses in systems analysis and design and database systems. His research interests include IT-enabled organizational change, virtual teamwork, electronic commerce, IS failures, and qualitative research methodologies. Suprateek can be reached by e-mail at sarkers@cbe.wsu.edu.

Allen S. Lee is a professor in the Department of Information Systems at Virginia Commonwealth University and the Eminent Scholar of the Information Systems Research Institute. He has been a proponent of qualitative, interpretive, and case research in the study of information technology in organizations. He is Editor-in-Chief of *MIS Quarterly* and co-editor of the book, *Information Systems and Qualitative Research* (London: Chapman & Hall, 1997). Allen can be reached by e-mail at AllenSLee@csi.com.

Appendix
MANCO and TELECO Case Summaries

ORGANIZATION CHARACTERISTICS	MANCO	TELECO
Age as of 1996	30 years approximately	150 years approximately
Industry	Air purification equipment	Telecommunications
Size (before reengineering)	250 approximately	3,000 approximately
Culture (before reengineering)	Fragmented, inter-functional hostility, politically charged, task-oriented, narrow compartmentalized thinking, sluggish action	Monopolistic, technology-driven rather than customer-driven thinking, inter-functional indifference, unionized, job security assumed
Size/Revenue (after reengineering)	Same headcount; Revenue increased considerably	Headcount reduced by 800 and then increased by 500 (net reduction 300). Revenue increased considerably.
Culture (after reengineering)	Agile, cheerful, cross-functional cooperation	Market-oriented, cross-functional, individualistic and transactional
REENGINEERING:		
The "definition-in-use" of reengineering	"Organizational reform" for excellence *using common-sense and IT*	Downsizing and reorganizing using IT and principles such as collocation, hand-off reductions, etc.
Reason for reengineering	To avoid "extended mediocrity"	Very hostile competitive environment as a result of changes in regulations
Goals of the initiative	To excel and take advantage of market opportunities	To survive in a competitive environment after many years in a monopolistic environment.
Nature of the reengineering process	Radical, structural reorganization, followed by IT-enabled process change, followed by incremental adjustments of the social organization and technology	Simultaneous radical change is the social organization and the organizational information technology.
Sample computerized tools used in business process redesign	Easy Flow	Visio and Project

PROCESS REDESIGN:		
Nature of the redesign process	Autocratic changes in the structure, followed by participative, iterative redesign	Empowered redesign of business processes through an understanding of organizational needs within staffing constraints set unilaterally by the top management. The overall process was very stressful.
Nature of the vision (redesign)	*Organizational agility* was the broad vision articulated by the top management; vision for specific processes evolved through interactions of different social/functional options with different technical options. The redesigned processes were not formally represented on paper, but shared in the minds of the team and close associates.	Bottom-up vision was created by the redesign team. Constraints such as the number of people to be retained and total number of processes set by top management.
Primary role of top management in the redesign	Created structural and cultural context for effective cross-functional processes; complete support provided to reengineering team throughout	Committed primarily to the idea of personnel reduction; specified the number of business processes in the "redesigned" organization; provided all deadlines
Role of the redesign team	To discover the organization and redesign suitable for the company. Also responsible for implementation	Creating and selling the vision; not given the power to implement.
Role of IT envisaged	Providing a set of tools, accelerated information sharing, detailed management information	Substituting employees through automation and information sharing.
Role of IT tools	Limited; use of tools discontinued during redesign	Considerable use in documenting the processes (before and after), project schedules, etc.
Whether redesign was seen by stakeholders as effective	Yes; though one department saw one aspect of the vision as unrealistic	Yes, most stakeholders felt that the redesign was reasonably effective

IMPLEMENTATION:		
Nature of the implementation process	Extremely planned; 3 pilots	Use of the "parking lot" strategy made implementation very stressful; every organizational member was relieved of his/her position and some were rehired into new positions in the redesigned organization.
Nature of communication	Superficial (sometimes misleading) formal communication	Some formal communication explaining the changes and impacts; no communication allowed during redesign
Nature of IT implementation management	Very systematic	Ineffective in managing relationships with IT vendors; IT lead-times grossly underestimated.
Nature of pre-existing IT infrastructure	Poor	Somewhat better
Role of top management	Complete support; senior VP had hands-on involvement	Very hands-off approach; commitments changed over the implementation phase.
Main problems faced in implementation	Moving to a more sophisticated IT infrastructure	IT infrastructure inadequate; IT and Human Resource action completely uncoordinated—IT not delivered on schedule but personnel laid off as planned earlier, resulting in almost a shutdown of customer support operations.
Morale during implementation	High overall	Fluctuating—often very low.
Degree to which the "redesign" was implemented	Fairly large	Quite small
Definition of success (before the initiative was undertaken)	Cross-functional integration, creation of useable information for effective management	20% reduction in work-force while improving service
Whether the implementation was seen as successful by stakeholders	Yes	No, most didn't.

14 UNDERSTANDING E-COMMERCE THROUGH GENRE THEORY: THE CASE OF THE CAR-BUYING PROCESS

Sue Conger
Ulrike Schultze[1]
Southern Methodist University
U.S.A.

Abstract

There is a dearth of theory-based guidance for organizations making decisions about electronic commerce and about enhancing their business processes through the use of new media, particularly the World Wide Web (Web). Genre theory provides a useful scaffold for generating such guidance. Communication genres are generic responses to recurring situations. Examples include a business letter, a meeting or a memo (Yates and Orlikowski 1992). In this paper, we apply genre theory to evaluate the business processes involved in buying a car. From this analysis, we hypothesize changes in the business process and in buyer/seller satisfaction that result from electronically mediated genre of communication.

Keywords: Gene theory, automotive industry, electronic commerce, communications, richness theory, car buying process.

1. Introduction

More and more organizations are engaging in electronic commerce to cut costs and stay competitive (Girishankar 1998). Furthermore, some services that did not exist previously now form the basis of viable on-line or "virtual" businesses. In many cases,

---

[1]Authors' names are in alphabetical order. Both contributed equally to the ideas presented in this paper.

however, migrating business processes to the Web has not resulted in the kind of business success that some predicted. For instance, Amazon.com is still not profitable despite a phenomenal growth in sales during 1998.[2] The airlines are trying to entice customers with incentives for electronically-booked flights (e.g., Travelocity, Delta).

Genre theory is useful in explaining the mixed success of e-commerce because it provides a scaffold for thinking through the implications of adding new media to the existing media repertoire. Generic communicative responses, or genres, develop in response to events that recur in a situated social setting (Orlikowski and Yates 1994). For instance, a letter of recommendation is a response to a potential employer's desire to hear a previous employer's evaluation of a candidate. A letter of recommendation represents a genre recognizable primarily by its *purpose*, i.e., the socially constructed understanding of the intent, content and expected use of the communication (Miller 1984). There are other genres, such as an e-mail, that are primarily recognizable by their *form*. Form is an amalgamation of structure, medium, and language (Yates and Orlikowski 1992).

In this paper we integrate genre theory with business process thinking to explore the impact of the Web on the car buying process. The car buying process is a series of communicative acts which are accomplished through a genre system, i.e., a set of interrelated genres that interact with each other in specific settings (Brazerman 1995). We have chosen the car industry because of the attention it has received with respect to e-commerce (Kichen 1997). Furthermore, the car industry is one of the largest (20 million units a year) and one of the most studied (Llosa and Lee 1998).

Because buying a new car is a substantial purchase and thus a significant decision for most people; the car buying process is somewhat stressful, requiring deliberate and "mindful" information processing (Kichen 1997). For many customers, especially women, it is an also an unpleasant experience because many sales associates assume that they know nothing about cars and treat them with condescension (Chisholm 1999; Mahoney 1991; Prochazka-Dahl 1997). Furthermore, the used-car-salesman stereotype, that is, a "car guy with a beer gut and greased back hair, wearing polyester pants and a plaid jacket"[3] who is ready to sell unsuspecting customers a "lemon," negatively impacts car sales associates in general. The disintermediation of Web technology and the provision of vast amounts of information should have a positive effect on car buyers in general, and women car buyers in particular (www.womanmotorist.com).

Recent statistics show that about 20% of new car buyers, about four million per year, now research their purchases on the Web (Black Enterprise). As this number grows, changes in the car purchase process are becoming clear. Genre theory provides the lens for exploring those changes and drawing conclusions with both theoretical and practical value.

Five steps, or communicative acts, comprise the car buying process. These are (1) information gathering, (2) test driving, (3) qualifying the buyer, (4) negotiating the price, and (5) finalizing the deal. For each communicative act, we explore three broad

---

[2]In 1998, Amazon.com posted book sales of $610 million; a 313% increase over 1997 sales (*New York Times Magazine*, 3/14/1999).

[3]Paraphrased from an interview with a regional retail development manager at a large U.S. car manufacturer (4/1/1999).

categories of genre—face-to-face, paper-based, and Web-enabled. We include e-mail, web pages and electronic discussions in our Web-enabled category. While our three genre categories may be an oversimplification, this level of analysis allows us to develop a coherent set of hypotheses about an entire business process. We ask questions like: What happens when the communicative act of price negotiation becomes computer-mediated? How does each communicative act change? How is the sales/buying process as a whole affected? What do these changes mean for buyer and sales associate satisfaction?

Our discussion proceeds as follows. We outline genre theory and develop a theoretical framework that integrates genre theory with business process thinking. Then we turn our attention to the automotive industry, and describe the new car sales process in terms of its communicative acts. For each communicative act, we explore the different genres and hypothesize what impact Web-enabled genres will have on the car buying process, particularly with respect to such outcome variables as cost, customer satisfaction, sales associate satisfaction, and the time it takes to close a sale.

2. Genre Theory

Genre is a concept derived from literary theory that distinguishes different forms of literature, including the poem, the drama, the tragedy, the comedy, and the novel (Abrams 1988). Recently, the concept of genre has been expanded to describe business communications, such as the business letter, the memo, and the e-mail message (Orlikowski and Yates 1994; Yates and Orlikowski 1992). Genre is generally defined in terms of content and form, which are further broken down into the following features:

- **Substance**: the purpose and content of the communication, e.g., a question.
- **Structure**: the format in which information is presented, e.g., the "fill-in-the-blank" form.
- **Media**: the medium through which the communication occurs, e.g., face-to-face or electronically mediated.
- **Language**: the type of expression used in the communication, e.g., jargon or formal language.

These features of genre are highly interdependent and dynamically intertwined (see Figure 1), implying that a change in one genre element will cause a ripple effect through the other elements. The work by Orlikowski and Yates (1994) on the development of the e-mail genre from the business letter illustrates the events that trigger genre changes and the kinds of adjustments made in genre elements to accommodate social and technological changes.

Shepherd and Watters (1999) argue that the use of the computer and the Internet has made it necessary to extend the definition of genre as a combination of content and form. They propose a new class of genre, namely, "cybergenre," which is defined in terms of content, form, and functionality. Functionality refers to the capability of the new medium, e.g., interactivity through search engines.

Since our objective in this research is to hypothesize the impact of "cybergenres" on the car buying process, we need to compare these new genres with existing ones. We therefore apply Orlikowki and Yates' more generic definition of genre.

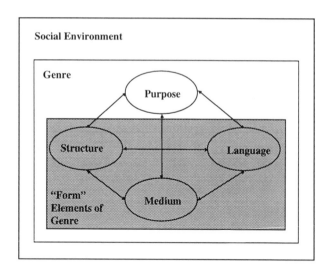

Figure 1. Dynamic Relationship between the Features of Genre

In this paper, we focus on the effect of a technology-induced medium shift on the communicative acts that compose the new car buying process. Media channels embody specific features such as anonymity and asynchronicity (Griffith and Northcraft 1994). Through the identification of these features and the tracing of their implications for the four genre elements, we explore the effect of a new medium on each communicative act independently and on the overall business process.

2.1 Theoretical Framework: Business Processes as Genre Systems

By providing the theoretical scaffold for tracing the ripple effects of a medium change through the other genre elements to a change in communicative and, ultimately, social practice, genre theory is a powerful tool for analyzing the implications of introducing new media into a business process. Business processes are a series of communicative acts that can be accomplished through a variety of socially recognizable responses, i.e., genres. Depending on personal preferences and contextual circumstances, a specific genre or set of genres can be selected to achieve a communicative act. The relationship of business processes to genres is depicted in Figure 2.

With this framework, we can generate hypotheses about the implications of adding a new medium to the mix of genres that make up a genre system. It is through such a genre system that a business process is accomplished. When a new medium, such as electronic communication via the Web, becomes a viable alternative to the existing media, we would expect a new category of genres to develop (i.e., a new horizontal band in Figure 2). Since the genres that are available to accomplish one communicative act

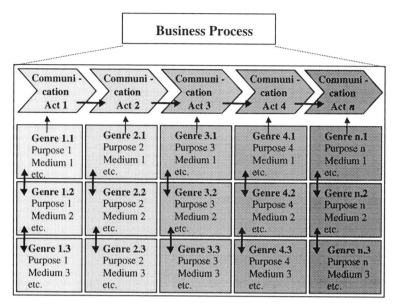

Figure 2. Business Process as a Genre System

in the business process are intertwined, the new Web-enabled genres[4] will induce changes in the genres belonging to the other genre categories (e.g., face-to-face and paper-mediated genres.) In Figure 2, these changes are depicted by arrows. The consequences of a new medium on both individual communicative acts and the business process as a whole are largely dependent on the features embodied in the medium (Griffith and Northcraft 1994). In genre theory, these media features find expression in the structure element. In the next section, we apply the model to the business process of buying a new car.

3. The Business Process of Buying a New Car

The car buying process includes the following communicative acts:
- *information gathering:* a communicative act in which car buyers are informed of their alternatives and narrow their purchasing choices;
- *test drive (or tryout):* a communicative act in which the car buyer interacts with the car directly;
- *qualification of a buyer:* a communicative act in which the sales person seeks to determine the buyer's willingness and ability to pay for the intended purchase;

---

[4]Shepherd and Watters (1999) identify the following Web-enabled genres: home page, brochure, resource, catalogue, search engine and game. Crowston and Williams (1999) define FAQs as a Web-enabled genre.

- *negotiation of price:* a communicative act in which the car deal is structured with respect to trade-ins and discounts, as well as financing; and
- *finalization of sale:* a communicative act in which payment is made and physical goods such as title documents, keys and cars are exchanged.

These communicative acts can overlap and occur simultaneously. For instance, the test drive may coincide with buyer qualification. We will now examine each of the communicative acts in turn to establish which communication genres apply to them and how Web-enabled genres affect them.

3.1 Information Gathering

The purpose of information gathering is for car manufacturers, dealers and other agencies to influence the customer's purchasing decision with respect to the make, model, and year of the car, as well as the choice of dealership with whom the customer will do business. For the buyers, the purpose of information gathering is to gain an understanding of the car market and to narrow their purchasing alternatives. The *content* of genres used for information gathering includes primarily the features, functions, pricing and options of the car, as well as assessments of the car. Appendix A lists examples of genres as well as the structural and language components that have developed in response to buyers' recurring information gathering needs.

A large number of communication genres have developed in response to the needs of dealers and manufacturers to influence the buyer's purchasing behavior as well as buyers' need to develop an understanding of car offerings. Manufacturers' and dealers' advertisements represent an advocacy-based informing strategy. Independent evaluations of cars range from the objective, such as *Consumer Reports*, to the car-enthusiast, such as *Car and Driver*, which represent a comparison-based informing strategy. These genres are instantiated across the three media groups used in this analysis. There are also several Web-enabled genres that are less focused. Bulletin boards, online discussion groups, and e-mails between correspondents also contain anecdotes and other informal information about cars.

It is in the *structure* aspect of genres that we see most of the differences between the three genre categories. Paper-based information gathering genres provide non-interactive, one-way, asynchronous communication between presenter and reader. The reader/buyer is unidentifiable and, therefore, the information presented is not customized to the buyer's needs or preferences. Information is distributed through many channels, including library, bookstore, and news agents.

Face-to-face information gathering communications are fully interactive, real-time, synchronous, and bi-directional between a sales associate and buyer. The buyer is thus known to the sales associate and information can be tailored to the buyer's requests and perceived needs. Typically, geographic proximity limits the variety of accessible information sources (e.g., sales associates.)

Web-enabled information gathering exhibits limited interactivity. Even though most Web pages afford only one-way communication, search capabilities provide buyers with more direct access to requested information. This potential for interactivity as well as the reduced need for space conservation (which is an especially challenging problem in print media), makes it possible for information providers to make more extensive and detailed information available to buyers. The electronic nature of the medium also reduces the cost of editing the information. This implies that buyers using Web-enabled genres have access to more accurate and up-to-date information than they do in print-mediated genres and more permanent access to information than they do in the genres mediated by face-to-face interactions. The degree to which the information is customized to buyers' needs and preferences in web-enabled genres depends on the specific genre they use and on whether they choose to be identified. For instance, in a car features site like www.edmunds.com the information is not customized and the buyer remains unidentified. However, if the car buyer were to e-mail a sales associate or a friend, they would be identified and the information gathered would likely be specific to the buyer's needs and preferences.

Finally, the *language* differs for each genre category, ranging from the formal and explicit in print media genres to the more informal and non-verbal in genres based on face-to-face interactions. In Web-mediated genres, the language is primarily formal and explicit into which are mixed many affective, non-verbal language elements (i.e., multimedia including graphics, photographs, audio and video). This suggests that web-enabled genres support rich communication (Daft and Lengel 1986).

Based on the differences between the genre elements of the face-to-face, paper-mediated and Web-enabled genre categories, we expect to see a number of changes in the information gathering stage of the car buying process as more and more buyers rely on Web-enabled genres (e.g., dealer and manufacturer Web sites and e-mail interactions with sales associates). First, Web-enabled genres will reduces the cost of information gathering for the buyer. In part, this is due to the asynchronous nature of the Web-enabled genres. Asynchronicity reduces the coordination costs that are typically incurred in face-to-face environments (Kambil and van Heck 1998). The cost savings to the buyer can further be attributed to the fact that buyers can access multiple sources of information through one channel, that is, the computer. This saves travel costs related to obtaining print media from bookstores and libraries, for instance. The time spent gathering information should be reduced when Web-enabled genres are used because these genres support some degree of interactivity, e.g., search facilities reduce the time it takes to locate specific pieces of information.

Second, because the cost of information gathering is lower for electronic media, buyers, per period of time, will be able to gather larger amounts and a greater variety of information than they would if they had relied on face-to-face or paper-mediated genres (Kanell 1998, Kichen 1997). Because car purchases are infrequent and substantial enough for consumers to value information highly (Dunnan 1998), we expect that the car buyers will spend the same amount of time gathering information irrespective of medium.

Third, buyer satisfaction with the information gathering genre is a function of time, cost, perceived value and availability of information. Buyer satisfaction should increase with the use of Web-enabled genres because of lower costs and greater information availability. With the increase of information, buyers will feel more informed and more confident and satisfied with their decisions (O'Reilley 1980).

Furthermore, buyers' ability to choose whether they want to be identified by the information providers during their information gathering activities is also expected to increase buyers' satisfaction. The anonymity of electronic media should be particularly relevant to women who generally dislike dealing with a car sales associate because of the condescending attitudes that male sales associates display toward them (Chisholm 1999, Prochazka-Dahl 1997). Web-enabled information gathering proceeds in an impersonal buyer-paced context, in which the buyer remains unidentified and less exposed to pressures pushing him/her into a premature purchasing decision. Thus, anonymity and asynchronicity, both of which are structural features of Web-enabled genres, are expected to contribute to higher buyer satisfaction (Kichen 1997).

Fourth, with the increased use of Web-enabled genres on the part of buyers, the sales associate's job of influencing a buyer's behavior becomes more difficult. Consequently, the sales associate's satisfaction with the information gathering stage of the car buying process is expected to decline. In part, this is because influencing an informed buyer is more difficult than influencing an uninformed buyer given that the latter has little or no basis for assessing the validity of the sales associate's claims. Furthermore, the more informed buyer is likely to be the buyer that already mistrusts car sales associates.

Another challenge for the sales associate lies in influencing a misinformed buyer. Buyers relying on Web-enabled genres are likely to feel highly informed (given the supposed accuracy and up-to-date nature of the information available through the Web[5]), and therefore particularly suspicious of a sales associate's effort to prove them wrong. The number of misinformed buyers is likely to increase with the use of Web-enabled genres because of information overload on the part of the buyers. When a buyer is in a situation of information overload, i.e., a condition in which the amount of information that warrants attention exceeds the individual's ability to process it (Schultze and Vandenbosch 1998), he/she may feel more informed and more satisfied with the information gathering activity. However, because the information available exceeds the buyer's processing capacity, he/she is incapable of separating the relevant from the irrelevant, which leads to poor comprehension and decision making (O'Reilley 1980).

Based on the above discussion, we can formulate the following hypotheses related to the use of Web-enabled genres for the communicative act of information gathering:

H1.1:   When Web-enabled genres are used in information gathering, the buyer's perceived cost will be lower than when genres based on other communication media are used.

H1.2:   Buyers will gather more and a greater variety of information when they rely on Web-enabled genres than they do when they rely on genres based on other communication media.

H1.3:   Buyers' satisfaction with information gathering will be higher when they rely on Web-enabled genres than when they rely on genres based on other communication media.

H1.4:   The sales associate satisfaction with information gathering will decline as more buyers complete their information gathering activities through Web-enabled genres.

---

[5]Information on the Web is regarded as accurate and reliable until the user finds out otherwise (Chisholm 1999; Kanell 1998).

3.2 Test Drive

The purpose of a test drive is for the buyer to interact with a car. The primary participants in this communicative act are thus the buyer and the car. However, a sales associate typically accompanies the buyer on a test drive not only to control the duration and route, but also to highlight features of the car and answer the buyer's questions. The test drive thus overlaps somewhat with the communicative act of information gathering. The *content* of test drive genres is focused primarily on the features of the car, both in explicit as well as sensory and non-verbal terms.

Paper-based genres are not viable for this stage of the car buying process and Web-enabled genres offer only a weak substitute for the face-to-face experience. This is because the tactile, experiential nature of the interaction between the car and buyer is only partially emulated in the Web-enabled genres. Appendix B summarizes the comparison between the three genre categories for the communicative act of test driving.

The face-to-face test drive is *structured* in the following way. The sales associate follows a protocol of preparation during which the buyer's driver's license and proof of insurance is procured. This is followed by a 10 to 15 minute test drive. The buyer is fully identified. The communication is interactive and customized to the buyer. The sales associate has the power to deny a test drive and may determine its route and duration.

The *language* between car and buyer/driver is non-verbal, tactile and affective. The language between sales associate and buyer that occurs during a test drive is informal and typically focused on features of the car.

In Web-enabled genres, a virtual test drive supported by virtual reality or Quicktime™ provides an unlimited, user-directed and interactive experience through makes and models. The buyer remains anonymous. Thus, the *structure* of the Web-enabled test drive varies considerably from that in the real world.

Similarly, the *language* in the Web-enabled test drive genre is quite different from that in the face-to-face genre. The Web-enabled genre uses non-verbal language including flashpix, video, and images, as well as explicit, verbal language in the form of text and voice. For instance, the virtual experience is accompanied by persuasive marketing commentary, which describes what the buyer would perceive in the real world. However, the sensory nature of the interaction between the car and the driver cannot be recreated over the Web at this time.

In summary, we do not expect Web-enabled genres to have a significant impact on the communicative act of test driving but virtual test drives may help buyers narrow their car choice. This will require buyers to test drive fewer cars than they might otherwise have done. This suggests that virtual test drives speed up the car buying process as a whole and they reduce the amount of time a sales associate interacts with the car buyer. We derive the following hypotheses for the communicative act of test driving:

H2.1: Web-enabled genres used for test driving, e.g., virtual test drives, will not replace "real" test drives.

H2.2: The use of Web-enabled genres for test driving will reduce the number of real test drives that a buyer will want to make.

3.3 Qualification of a Buyer

The purpose of this communicative act is for the sales associate to determine how serious the buyer is with respect to purchasing a car and, more specifically, a car from the sales associate's dealership. This assessment includes a determination of the buyer's ability to pay for the make and model of car in question. In a face to face context, the sales associate typically relies on social cues such as dress, age, ethnicity and general demeanor to make such an assessment. The associate seeks to elicit information on the customer's profession, the time frame within which they expect to make the purchasing decision, and other dealerships under consideration.

Buyer qualification can be accomplished though a variety of genres, although those based on face-to-face interaction and electronic media are the most prevalent. (The differences between the three genre categories are summarized in Appendix C.) In the case of paper-based genres, the *structure* of the communicative act is interactive and asynchronous and most of the interactions will be conducted through the mailing of paper forms. The *language* on these is formal and explicit, and the forms are likely to consist of a fair amount of legalese.

When the face-to-face genre is used for buyer qualification, the sales associates are gathering qualifying information throughout their interactions with the buyer. This starts at the initial point of contact with the buyer. Buyer qualification is done in an informal manner with the sales associate observing buyers and asking them a series of questions. The *structure* of face-to-face genres is thus interactive, synchronous and primarily initiated by the sales associate. The *language* of buyer qualification is informal and, to some extent, non-verbal.

Purchasers relying on Web-enabled genres have the option of communicating directly with dealers via e-mail, or being referred to dealers through third parties, e.g., Auto-by-tel. The e-mails typically request an appointment to test drive a car or to evaluate a trade-in car. By initiating this process through Web-enabled genres, a buyer is deemed to signal seriousness.[6] Given the novelty of Web-enabled genres in the car buying process, it may be reasonable for sales associates to assume that the use of e-mail to schedule an appointment signals a customer's seriousness. With more widespread use of this genre, we would expect its value as a proxy for buyer qualification to deteriorate. For instance, buyers will schedule appointments via e-mail that they then do not keep.

Buyers can be pre-approved for credit through banks or other lending institutions. Credit applications can be made over the Web, and dealers who are strategically aligned with the lending institution at which the buyer files an application would have knowledge of the customer's ability to pay for the car.

Buyers can gain a better understanding of what constitutes a legitimate qualification procedure (i.e., what questions and information requests are reasonable) when these are published explicitly on dealers' Web sites or compiled into forms. This creates the impression that the communicative act of buyer qualification is less subjective and personal. Buyers will thus feel less vulnerable and exposed (Kanell 1998, Mahoney 1991). The formality of the *language* and the explicitness of information play an

---

[6] This insight was gained through personal correspondence with a sales associate in a car dealership based in a large city in the southwestern region of the U.S. (1/15/99).

important role in increasing the buyer's satisfaction with the buyer qualification communicative act.

Furthermore, the asynchronously *structured* interaction with the dealership/sales associate implies that the buyer has time to contemplate the information that he/she is asked to provide. Thus, when Web-enabled genres are used, the buyer has more control over the timing of this communicative act than he/she does in face-to-face genres. We expect the level of satisfaction to increase more among women than among men because women feel particularly uncomfortable being questioned and pressured by sales associates who are primarily male (Mahoney 1991, Prochazka-Dahl 1997).

By making the buyer qualification procedure more explicit through Web-enabled genres, sales associates lose some of their knowledge advantage over buyers This loss of power is likely to lower sales associates' satisfaction. The use of Web-enabled genres will also force sales associates to change their work practices. In order to meet Internet users' expectations, sales associates need to respond to customers within 15 minutes (Conger and Mason 1998). There is also an increased need to manage an appointment schedule. This increased job pressure, coupled with a loss of power over the buyers, is likely to decrease sales associates' satisfaction.

The hypotheses relating to buyer qualification are:

H3.1: The use of Web-enabled genres during buyer qualification will increase buyer satisfaction.

H3.2: The use of Web-enabled genres during buyer qualification will decrease sales associates' satisfaction.

3.4 Negotiation of Price

The purpose of price and terms negotiation is to arrive at a mutually satisfactory deal. The minimum *content* of genres used in price negotiations includes offer terms and amounts, problems that the buyer might have in pursuing a purchase elsewhere (e.g., "You will not be able to get these terms at *xyz* dealership because they are exclusive here"), and statements that describe the quality of the deal (e.g., "You really are getting a bargain here"). Whether issues concerning "hold backs," dealer incentives, manufacturer's suggested retail price (MSRP) and its components make it into the communicative act of price negotiation depends on the buyer's prior knowledge of these. Buyers, in general, distrust car dealers and have widespread beliefs that dealers withhold information about vehicle quality and potential negotiating points. The price negotiation phase of the car buying process represents a time when the stakes are high for both buyers and sales associates. It is thus also a time of tension when emotions run high.

While all three categories of genres can be used to conduct price negotiation, paper-based correspondence is infrequent for reasons similar to those discussed under the "buyer qualification" section. In the interest of space, is not discussed further here. Please refer to Appendix D for a summary of the difference between the three genre categories as they apply to price negotiation.

To help sales associates *structure* price negotiations in their favor when face-to-face genres are used, some car retailers train them to control the exchanges with buyers during this phase of the car buying process. They use scripts that avoid topics such as "hold backs." This implies that sales associates are better prepared with the generic responses required during price negotiation than the buyer is in the face-to-face environment.

Web-enabled genres provide buyers with more information about how deals are structured, what the price components are[7] and how prices compare across dealerships. A buyer equipped with detailed price information can exert pressure on a sales associate to account for all price components and negotiate with respect to each (Kichen 1997). The margins that sales associates will be able to make on the sale of a car are likely to shrink as a result of the use of Web-enabled genres. Furthermore, buyers can find tips on how to negotiate with sales associates on the Web. This gives buyers a greater degree of control over price negotiation.

In the case of price negotiation, the asynchronous *structure* of communicative acts supported by Web-enabled genres decreases the sales associates' ability to put pressure on buyers to make their decision quickly. The use of Web-enabled genres gives buyers the opportunity to negotiate deals with multiple sales associates at the same time. Buyers therefore have more comparative information and more time to evaluate a specific deal, making them feel more in control and less pressured to close on a deal. When Web-enabled genres are used during price negotiation, the buyer controls the communicative act more than the sales associate does.

The following hypotheses summarize the effects of using Web-enabled genres during price negotiation:

H4.1: The use of Web-enabled genres during price negotiation will increase buyer satisfaction.

H4.2: The use of Web-enabled genres during price negotiation will be more controlled by the buyer than by the sales associate.

H4.3: The sales associate's satisfaction with price negotiation will decline the more negotiation is with buyers educated through Web-enabled genres.

3.5   Finalizing Sale

The purpose of finalizing the sale is to complete terms and all necessary paperwork, to exchange physical items such as documents of ownership, keys, and cars, and to conduct a final briefing on the features of the car. The content of this communicative act includes title and registration for any trade-ins, buyer's insurance verification and driver's license, loan documents for signatures and exchange, and various required documents about odometer readings and car condition. Face-to-face genres are more likely than other genre categories to include assurances of the quality of the deal and of the vehicle to delay the onset of buyer's remorse.

The use of paper-based genres for finalizing the sales is possible due to the physical nature of the artifacts that need to be exchanged between the buyer and the dealership, e.g., car titles, checks and keys. The electronic nature of Web-enabled genres does not support the transfer of physical artifacts and is, therefore, of use only in some aspects of the sales finalization stage of the car buying process.

---

[7]Examples of web sites with information about price components: www.uniprimeinc.com, www.carpoint.com, www.edmunds.com, www.intellichoice.com; www.autoreach.com.

The *structure* of sales finalization using face-to-face genres is determined by the finance person responsible for ensuring the completion of the appropriate documents. The role of the sales associate is one of reminding the buyer of the quality of the deal and of the vehicle. The *language* is formal and pertains mainly to the documents involved.

The major difference between the face-to-face and Web-enabled genres is the asynchronous structure of the Web-enabled communicative act. This implies that the elapsed time will be longer than if face-to-face genres are used, while applied time should be the same as in face-to-face genres. Applied time may, however, be shorter because electronic forms and document templates may be used to standardize the information exchange and thereby streamline the activities that make up the sales finalization stage of the car buying process.

Most buyers find parting with their money, committing to new car payments, and having their financial status discussed an uncomfortable experience (Chisholm 1999). As a result, any disintermediation that renders the exchanges required for the sales finalization less personal is expected to increase buyer satisfaction, especially among women.

The hypotheses generated from this discussion are:

H5.1: The use of Web-enabled genres during finalizing a sale will increase buyer satisfaction, especially among women customers.

H5.2: The use of Web-enabled genres during finalizing a sale will increase elapsed time of the communicative act while applied time will be no longer and may be shorter.

3.6 Overall Impact on the Car Buying Process

The increased satisfaction among car buyers that accompanies the use of Web-enabled genres is based on more cost-effective access to information and the ability to access more information about cars and about the price components. By becoming a more informed customer, car buyers can exercise more control over the car buying process. Furthermore, the buyer's ability to negotiate deals with multiple dealerships at the same time and thereby find the best deal, increases buyer satisfaction. The disintermediating and depersonalizing roles Web-enabled genres play in the car buying process are particularly valued by women customers, who are frequently treated with condescension by primarily male sales associates.

The asynchronous nature of Web-enabled communication suggests that there will be delays within communicative acts and between the different communicative acts. This increases the elapsed time of the entire car buying process. However, time saved by fewer test drives and more streamlined procedures during sales finalization is expected to reduce or keep constant the actual applied time it takes to complete the car buying process.

Car sales associates are under siege and there is pressure from car buyers and information providers to change their practices (Llosa and Lee 1998). For instance, with the increased use of Web-enabled genres, sales associates will have to surf the Web, use e-mail and schedule and keep appointments. Sales associates' satisfaction with their jobs is expected to decrease as more buyers become more knowledgeable about vehicles and prices through Web-enabled genres. Armed with more information, buyers can negotiate more aggressively and have more control over the communicative acts. Influencing buyers' decisions, especially when buyers are misinformed, becomes more of a challenge. The asynchronous nature of Web-enabled genres also decreases the sales associate's ability to put pressure on the buyer to make decisions quickly.

The following hypotheses summarize this discussion of the effects of Web-enabled genres on the car buying process as a whole.

H6.1: The use of Web-enabled genres will increase buyer satisfaction for the overall car buying process, especially for women customers.

H6.2: The use of Web-enabled genres during the purchase of a car will increase elapsed time of the process while actual applied time will be no longer and may be shorter.

H6.3: The use of Web-enabled genres will decrease sales associates' satisfaction with the overall car buying process.

4. Conclusion

This paper contributes to research on electronic commerce by developing a framework that combines genre theory and business process thinking. This is a general framework that can be applied to business processes other than the car buying process, on which we focus here. This paper generates a set of testable hypotheses that are conducive for future empirical research on the effect of the Web on the car buying process.

In its current hypothesis formulation phase, this research also contributes to practice by developing a framework for understanding the nature of the car buying process and the changes on this process by Web-enabled genres. This understanding is expected to help car dealers plan for the impacts of electronic commerce on their business by adapting, among other things, their hiring and training practices. We also anticipate that this framework will serve as a map for dealers who want to identify how they can differentiate themselves from other dealerships by adding value to a process that is becoming increasingly generic through the use of Web-enabled genres.

References

Abrams, M. H. *A Glossary of Literary Terms*, 5th ed. Orlando, FL: Holt, Rinehart and Winston, 1988.

Bazerman, C. "Systems of Genres and the Enactment of Social Intentions," in *Genre and the New Rhetoric.*, A. Freedman and P. Medway, (eds.). London: Taylor and Francis, 1995, pp. 79-101.

Business Wire. "Auto-By-Tel First Company to Complete Internet Car Buying Equation with Online After Market Sales." February 1, 1998.

Chisholm, P., and Bergman, B. "Personal Finance: The Tough Lot of a Car Buyer," *Maclean's*, January 11, 1999, p. 49.

Conger, S., and Mason, R. O. *Planning and Designing Effective Web Sites.* Boston: Course Technology, Inc., 1998.

Crowston, K., and Williams, M. "The Effects of Linking on Genres of Web Documents," *Proceedings of the Thirty-second Hawaii International Conference on System Sciences: Minitrack on Genre in Digital Documents*, 1999.

Daft, R. L., and Lengel, R. H. "Organizational Information Requirements, Media Richness and Structural Design," *Management Science* (32:5), 1986, pp. 554-571.

Dunnan, N. "Financial Fix: What's Yyour Investing Goal?" *Enter Magazine*, April 1, 1998.

Essence. "Road Scholars" (28), November 1997, pp 160-165.

Llosa, L. F., and Lee, E. Y.-J. "Car Buyer's Guide 1998: Sticker Shock! For the First Time in 20 Years, Prices are Falling—Yes, Falling. Here's How to Get the Best Deals on New Cars, Vans and Sport Utilities," *Money*, January 1998, p. 124.

Girishankar, S. "E-Commerce: Virtual Markets Create New Roles For Distributors," *Internet-Week*, April 6, 1998.

Griffith, T. L., and Northcraft, G. B. "Distinguishing between the Forest and the Trees: Media, Features, and Methodology in Electronic Communication Research," *Organization Science* (5:2), 1994, pp. 272-85.

Kambil, A., and van Heck, E. "Reengineering the Dutch Flower Auctions: A Framework for Analyzing Exchange Organizations," *Information Systems Research* (9:1), 1998, pp. 1-19.

Kanell, M. E. "AutoConnect Takes to the Net: Web Site Aims to Simplify Sales," *The Atlanta Journal and Constitution*, May 1998, p. B08.

Kichen, S. "Cruising the Internet," *Forbes Magazine*, March 1997, pp 198.

Mahony, R. "Car Buying: Why Women Get a Lemon of a Deal. (Discrimination in Automobile Selling)," *Ms. Magazine* (1:4), January-February 1991, p 86-88.

Maugh III, T. H. "Doctors Interrupt Too Quickly, Listen Too Little," *Los Angeles Times* Home Edition Health, January 25, 1999, p 4.

Miller, C. "Genre as Social Action," *Quarterly Journal of Speech* (70), 1984, pp 151-176.

Orlikowski, W., and Yates, J. "Genre Repertoire: The Structuring of Communicative Practices in Organizations," *Administrative Science Quarterly* (39), 1994, pp. 541-574.

O'Reilly, C. A. "Individuals and Information Overload in Organizations: Is More Necessarily Better?" *Academy of Management Journal* (23), 1980, pp. 684-696.

Prochazka-Dahl, L. "They Really Are Different (Selling Cars to Women)," *Ward's Dealer Business*, February 1997, p. 31.

Schultze, U., and Vandenbosch, B. "Information Overload in a Groupware Environment: Now You See It, Now You Don't," *Journal of Organizational Computing and Electronic Commerce* (8:2), 1998, pp.127-48.

Shepherd, M., and Watters, C. "The Functionality Attributes of Cybergenres," *Proceedings of the Thirty-second Hawaii International Conference on System Sciences: Minitrack on Genre in Digital Documents*, 1999.

Yates, J., and Orlikowski, W. "Genres of Organizational Communications: A Structurational Approach to Studying Communication and Media," *Academy of Management Review* (17:2), 1992, pp. 229-326.

About the Authors

Sue A. Conger has a Ph.D. from New York University in computer information systems. She has written two books, *Planning and Designing Effective Web Sites* (with Richard O. Mason) in 1998, and *The New Software Engineering* in 1994. Sue's research interests are electronic commerce, computer ethics, and software engineering. She is currently on the faculty at the Edwin L. Cox School of Business at Southern Methodist University in Dallas, Texas. She can be reached via e-mail at SConger@aol.com.

Ulrike Schultze is Assistant Professor in Information Systems and Operations Management at Southern Methodist University. Her research focuses on knowledge work, particularly the social processes of creating and using information in organizations. Ulrike has written on hard and soft information genres, information overload, knowledge management and the work practices that make up knowledge work. Her recent projects are in the areas of electronic commerce and virtual organizations. She received her bachelor's and master's degree in Management Information Systems from the University of the Witwatersrand in Johannesburg, South Africa, and her Ph.D. in MIS from Case Western Reserve University, Cleveland, Ohio. She can be reached at uschultz@mail.cox.smu.edu.

Appendix A: Genres for Information Gathering

Medium	Paper	Face to face	Web-enabled
Examples	- *Consumer Reports* - Manufacturer brochures - *Car and Driver Magazine* - Newspapers	- Speaking to sales person at dealership, either in person or over the phone - Speaking to friends, colleagues and trusted car mechanics to get recommendations	- Dealer Web sites: www.sewell.co - Manufacturer Web sites: www.lexus.co - Hobby Web sites on cars - Print media analogues: www.edmunds.co - Bulletin Boards - Online Discussions - E-mail with friends, sales person, etc.
Structure	Non-interactive; one-way communication between buyer and manufacturer, dealer or other information agency Asynchronous Buyer is unidentified; information not customized Information distributed across multiple channels, e.g., library, bookstore, newsagent	Fully interactive, bi-directional communication between sales associate and buyer Synchronous, real-time Buyer is identified; information customized Geography limits variety of accessible information sources	Limited interactivity; e.g., search capabilities promote direct access to information that will answer buyer's question Asynchronous Buyer can choose to be identified or not; consequently, information may or may not be customized Multiple channels available through one interface
Language	Formal, use of multimedia elements such as pictures	Informal, general business	Mix of formal and informal, contains multimedia elements such as pictures, video, sound.

Appendix B: Genres for Test Driving

Medium	Paper	Face to face	Web-enabled
Example	N/A	"Real" test drive	Virtual test drive using VR or Quicktime
Structure	N/A	10 to 15 minutes test drive after 10 to 30 minutes of protocol (e.g., taking driver's license)	Unlimited browsing through makes and models
		Buyer is identified	Buyer is anonymous
		Interactive (between buyer and car, and between buyer and sales associate)	Interactive (mouse driven)
		User-directed driving experience but sales associate has power to deny test drive and to determine route and duration of drive	User-directed
Language	N/A	Non-verbal, tactile, and experiential (between car and buyer)	Multi-media: video, falshpix images, text and voice
		Informal language (between salesperson and buyer)	Informal yet persuasive marketing language

Appendix C: Genres for Buyer Qualification

Medium	Paper	Face to face	Web-enabled
Examples	Pre-qualification through banks or dealers	Speaking to sales person in dealership	Pre-qualification for car loan through Web sites like www.uniprimeinc.com, www.carpoint.com Sending an e-mail (e.g., through dealer Web site) to make an appointment
Structure	Buyer-initiated contact; financial information is elicited via forms Interactive, asynchronous	Information for buyer qualification is gathered throughout information gathering and test drive phases in an informal way; i.e., through a surreptitious asking of questions Interactive, real-time	Buyer-initiated contact; buyer provides minimum information such as name and contact information; if more information is required this will be elicited through a form that the buyer is asked to fill out Interactive, asynchronous; some time pressure for dealership to respond to customer's request for an appointment in a timely manner
Language	Formal, structured forms to complete	Informal, conversational	Formal, structured forms to complete for credit applications. Appointment scheduling may be formal, via forms, or informal, via e-mail.

Appendix D: Genres for Negotiation of Price

Medium	Paper	Face to face	Web-enabled
Example	Correspondence between dealer and buyer through letters (rare)	Interaction between buyer, sales associate, and manager/ supervisor	Negotiation via e-mail between sales associate and buyer
Structure	High elapsed time, especially if buyer and dealer are not in agreement	Time varies but is less than other media	Timing a function of e-mail monitoring response by sales associates and buyer
	Asynchronous, limited time pressure; more thought time for both parties to deliberate over their prices and terms	Synchronous, some pressure for decision because negotiation is directed by sales associate; less thought time for buyer	Asynchronous, limited time pressure; more thought time for both parties to deliberate over their prices and terms
	Buyer may have two or more negotiations going on at once; opportunity for comparison	Limited opportunity for buyer comparison because of geographic distance between competing dealerships	Ease of buyer comparison (may have two negotiations going on at once)
	Buyer can be anonymous for part of the negotiation	Buyer is identified	Buyer can be anonymous for part of the negotiation
Language	Formal	Fairly formal (professional)	Mostly formal

Appendix E: Genres for Finalizing the Sale

Medium	Paper	Face to face	Web-enabled
Example	Car delivery and exchange of actual documents must be in person; remainder of process could be conducted through letter and/or fax	Interacting with sales person at dealership	Car delivery and exchange of actual documents must be in person; remainder of process could be conducted through e-mail
Structure	Sales associate, finance person and buyer direct process equally	Finance person directs the process; sales associate's role is to remind buyer of the quality of the price and vehicle	Sales associate, finance person and buyer direct process equally
	Asynchronous	Synchronous	Asynchronous
	Document turnover of title and registration; verification of insurance card and driver's license; and signatures for loan and/or checks, acceptance of car mileage, and acceptance of car condition	Document turnover of title and registration; verification of insurance card and driver's license; and signatures for loan and/or checks, acceptance of car mileage, and acceptance of car condition	Document turnover of title and registration; verification of insurance card and driver's license; and signatures for loan and/or checks, acceptance of car mileage, and acceptance of car condition
Language	Formal	Formal	Formal

15 BALANCING FLEXIBILITY AND COHERENCE: INFORMATION EXCHANGE IN A PAPER MACHINERY PROJECT

Helena Karsten
Kalle Lyytinen
University of Jyväskylä
Finland

Markku Hurskainen
Solution Garden, Inc.
Finland

Timo Koskelainen
Valmet Paper Machines
Printing Paper Machines
Finland

Abstract

The problem of balancing coherence and flexibility in collaborative information system design is approached here with two pairs of concepts. Boundary objects can support communication for perspective taking between communities of practice. Conscripting devices can support communication for perspective making within a community of practice. These theoretical lenses are used to study the uses of the technical specification in paper machine projects. Our study showed that as a boundary object it provided enough flexibility to allow negotiations, and sufficient local structure, for carrying out work in both communities of practice, the customer and the manufacturer. As a conscription device in the relatively virtual manufacturer project team, however, it proved to be problematic, due to the unidirectional nature of its construction, its diminishing importance in the web of conscripting objects, and its inconvenience as a means for learning. In search for balancing coherence and flexibility, the issues identified seemed to relate to acknowledging the dialectics of perspective making and perspective taking with boundary objects and conscription devices, to the openness and modifiability of these objects, and to the bounded transparency of these processes.

Keywords: CSCW, document management, coordination, boundary objects, conscription devices, perspective making, perspective taking, community of practice, designing collaborative information systems.

1. Introduction

A key issue in designing collaborative information systems is balancing coherence and flexibility (e.g., Wastell 1993). A coherent information system is capable of being a meaningful, familiar, and useful resource for different communities of practice (Lave and Wenger 1991), and it yields to formal specification to handle its complexity. A flexible information system is able to meet the local, situated activities and interests within a community of practice; it is robust in the face of all these uncertainties (on complexity vs uncertainty, see Mathiassen and Stage 1992). Due to the difficulty of this balancing act, collaborative information systems design also has remained more a craft than a careful application of a theory-based design methodology. The key problem in designing collaborative information systems is thus: if too much coherence is emphasized in the overall design (as many enterprise resource planning, or ERP, efforts do) at the cost of local flexibility, the system may face the problem of becoming an alien, possibly useless resource in actual work practice. If flexibility in the local design is emphasized, the overall design may lose coherence and become useless in its wider collaborative function. Where these meet, there may be considerable tension, for example, between uses of the "under-specified" document databases (local flexibility), and the "overly formalized," detailed entries in enterprise resource planning systems (overall coherence).

For examining this dilemma, we employ two related concept pairs: boundary object and conscription device, and perspective making and perspective taking. Star introduced the *boundary object* to be "an analytic concept of those (scientific) objects which both inhabit several intersecting social worlds and satisfy the informational requirements of each of them" (Star and Griesemer 1989, p. 393). Henderson has continued Star's work by introducing a specific kind of boundary object, the *conscription device* (Henderson 1991, 1998b), which "both enlists and constrains participation in creating and maintaining the object." The users of a conscription device must engage in inputting its elements and in revising them, if it is to serve its intended function for the users' purposes. These concepts have been used in analyzing why a certain technical solution has not been workable (e.g., Henderson 1991) or would not be workable (e.g., Mambrey and Robinson 1997).

Boland and Tenkasi (1995) have introduced perspective making and perspective taking for discussing how to represent and integrate knowledge across organizational units and boundaries. They refer to communication that strengthens the unique knowledge of a community as *perspective making*, and communication that improves its ability to take the knowledge of other communities into account as *perspective taking*. They tie perspective taking and boundary objects as follows: "Once a visible representation of an individual's knowledge is made available for analysis and communication, it becomes a boundary object and provides a basis for perspective taking" (Boland and Tenkasi 1995, p. 362). We continue from this and suggest a similar link between perspective making and conscription devices: When participants in a community engage in creating and maintaining a conscription device, they engage in perspective making, in communication that strengthens the unique knowledge of their community.

The main focus of our treatise is on establishing the value of these two concept pairs in tackling the coherence-flexibility dilemma in designing computer support for collaboration. We will use these interrelated concept pairs with our case to focus attention on issues that would need to be solved if a certain disorganized document set were to be redesigned to be a part of a collaborative information system, connected to data in ERP and other formalized systems. Our study will not give a detailed design to be implemented, nor detailed methodology rules to be followed, but it could be useful in pointing to a number of focal issues to be solved in the search for balance between coherence and flexibility.

2. Context and Method

Our case company is Valmet (http://www.valmet.com), one of the three largest paper machinery suppliers in the world. Recently, Valmet has restructured its business units to make them more independent from the specialized production units. There are tens of both of these, world-wide. In paper machinery delivery projects, the participating units need to coordinate, however, as "one Valmet," which presents a communication and informing problem. As a company level measure, Valmet has been reengineering its key processes to increase standardization and control throughout the production system, with the goal of cost and time savings. A key dilemma is now to tie together the increased informing and communicating to the increased discipline in production. In terms of information technology, these phenomena are visible in increased information sharing with electronic mail (in use for the past decade) and several thousand specific document databases in Lotus Notes (in use since 1992, with "100% coverage" since 1997), and a comprehensive enterprise resource planning system that is being implemented with Baan and related systems (in stages during the period 1997 to 2001). An aim for these initiatives has been to increase transparency of paper machinery projects and, as a consequence, increase efficiency within the whole of Valmet.

One of the sets of documentation to be overhauled is the *technical specification (TS)*, which specifies what kind of a paper machine is to be delivered from Valmet to the customer. While the TS is a set of documents, parts of the information in it—such as product configuration, cost calculation, and pricing—are managed by the ERP systems.

In this study we will give an account of its life cycle, and analyze it with the four concepts presented earlier, to assess how they could assist in solving the coherence vs flexibility dilemma. The practical purposes for our study are to contribute to redesign the structure and use processes of technical specifications, and to inform the choices of technology use.

This study was carried out in the Rautpohja units of Valmet in Jyväskylä, Finland, during 1997. Beforehand, two of us had accumulated considerable background knowledge by working as observing participants (Nandhakumar and Jones 1997; Taylor and Bogdan 1998) for nine months in a separate ISD project in the Project Department. Due to the complexity of paper machinery projects, this time was necessary and valuable for understanding project documentation. Plausibility, credibility, and relevant representation of our interpretations (Altheide and Johnson 1994) were further enhanced by our continued access to Valmet.

All paper machinery projects, and consequently the technical specifications of the paper machines, are different to a degree. Tracing the history and contents of more than one specific TS would have been a sizable effort, beyond the scope of one study. Therefore, the TS of a major current project, Project G, was chosen as the case for our study. The detailed progression of the TS document and its relations to other pertinent documents was traced, in Project G and in general, by interviewing 14 people, with eight representatives of the 50+ member large Project G team. Several descriptions were made (Yin 1989) of the process, of the documents, of communication practices around the TS in general, and of the breakdowns and disturbances encountered (Ngwenyama 1998) with the Project G technical specification. In the following description, each informant is identifiable by the number in parentheses. The interviews and observations made for another study are identified separately.

3. Building a Paper Machine

3.1 Paper Machinery Projects

Building a paper mill with a new paper machine is a major effort: a new, complete paper machine alone can cost from $100 million to $200 million. Due to this, the period from initial sales contacts to signing the contract can often be measured in years. But once the project has started, the start-up can take place in about 18 to 20 months. This is very compact, considering the size of the effort, allowing very little room for mistakes or delays, on either side. Even though there can be parallel and recursive chains of activities, three points of closure—signing the contract, reaching the freezing point for the design and, ultimately, the start-up of the machine—bring the processes together.

In Valmet, at each stage there are several departments and subcontractors involved. Due to the large size of the project, even though the *project manager (PM)* is the key mediator between the customer and Valmet, this boundary is kept rather leaky (Brown and Duguid 1998), with information going back and forth at all levels; that is, information that is "within the contract." This is also true within Valmet and toward the subcontractors. At the same time, the boundaries are impermeable, each Valmet unit guarding its technology and innovations with fervor. Even with this restriction, the amount of detailed information moving constantly between Valmet units and between the customer and Valmet is well beyond the scope of any one person. Therefore, the project is supposed to proceed along predefined paths. Only exceptions and major changes are brought to the attention of the PM.

3.2 The Technical Specification Documentation

The *technical specification (TS)* documentation in large projects can be 500 to 600 pages long, including texts, spreadsheets, and design drawings. It is relatively complex both in terms of structure and uses. During the sales phase, information or sub-bids are collected from the units and subcontractors. During the sales negotiations, there are often five to six cycles of revision before the customer is ready to start comparing the bids. The more knowledgeable the customer is, the more changes are likely to be required. In the actual contract negotiations, those pages of the TS to be included in the contract are initialed by both parties. Therefore the valid pages and information in them can only be confirmed from the paper document. By the time the project begins, the customer has approved of the main points of the technical specification. This does not mean that the technical specification is now approved and final: it will most likely be changed several times during the early stages of the project.

The project is started in a kick-off meeting, where the sales engineer tells how the sales process went and in what kind of spirit the deal was struck. The focal issues are what was sold, what was not sold, and what was sold with an option to add or remove. Dimensioning information are crucial, because they tell what size of machine is sold, how fast, and for what kind of paper with what kind of production. The head designers and their teams then pick out from the technical specification all pertinent information, and compare it to the current design guidelines that spell out the latest machine concept. The PM goes through the differences with the teams, ensuring that everybody understands the idea of the new machine.

During the project, the monthly two-day Project Meetings are the central events between Valmet and the customer, to clarify open issues, to make necessary changes, and to decide on additional investments. These are large meetings, with all the relevant people from both sides attending. Both parties may want to keep the TS as it is, because it is a part of the contract, and any major changes would need to be renegotiated in terms of price and delivery time as well. Therefore, it is likely that changes to the TS may no longer be written into it as revision pages, but that they are recorded in the meeting minutes, in acknowledgments of change orders, and as changes in job number lists, and conveyed by these to everybody in the project. By the start-up of the machine, the original technical specification is no longer valid but information in several other documents, taken together, describe in detail the paper machine that was built.

4. On Boundary Between Communities of Practice

Between Valmet and the customer, the technical specification document has different meanings and uses for both sides. As a boundary object (Star 1989; Star and Griesemer 1989), it is flexible enough to allow these interpretations, but in individual site use, it becomes strongly structured, setting the limits to what kind of machine Valmet will build and in what kind of site it is to be installed. As a whole, it translates the wishes of the customer to Valmet, and what Valmet is promising to build to the customer.

Each side needs a clear set of standardized methods (Star and Griesemer 1989, p. 393) by which their information is "disciplined" so that it could become a part of the translation, or, to use Boland and Tenkasi's concepts, each side would need to explicate their knowledge into a visible format in some way, available to the others, so that the other side can then comprehend and use it in their perspective taking. Examples of these methods are how issues are brought to the negotiation table as items on the agenda and how, only by initialing a page, the party becomes committed to the changes agreed upon during the negotiations. A series of boundary objects are generated, to maximize both the autonomy and communication between the camps. These boundary objects have been refined over decades of negotiations between machinery suppliers and buyers. During the sales phase, the subsequent versions of the technical specification and the meeting agendas and minutes, confirming the process, will become a "cascade" of boundary objects (Henderson 1991), culminating in a signed contract, with all pages of the technical specification approved by both sides.

As an agreement on the machine to be built, the technical specification contains the commitments from the customer to build the rest of the plant so that the machine fits in, and the commitments from the Valmet units that their machine fills its place. However, these commitments are never complete and never totally stable. The dilemma in the use of the technical specification is to work on something that is likely to be according to the commitment, but, at the same time, acknowledging the continuous need for negotiations, for repairs of breakdowns, with incomplete, possibly dated, information. This could be interpreted as intentional seeking of negotiation and acknowledgment of mutual influence (Robinson 1991) between the parties, when they repeatedly revisit issues until they are sufficiently well defined for both parties, yet leaving enough room for specific interpretations within a community of practice.

If the technical specification is looked at as a collection of sub-documents, these gain their near immutable quality when they leave the part of the organization where the knowledge of their meaning resides. After that, they might possibly be formatted by the sales assistants, but the content of the sub-documents is not changed by them or the sales engineers. These sub-documents now hold stable the intractable and heterogeneous materials from which they were composed, and which can now be conveyed, collated, and compared. However, the tailored quality of the paper machines dictates the need to modify the technical specification, and opens it up again for changes. From the viewpoint of the customer, the pages of the technical specification gain their immutable quality when they are initialed, when the changes have been approved by both sides. The chain of changes can be followed by tracing the versions of the technical specification. In Project G, this was achieved by retaining the whole history within the TS document, marked with asterisks. Thus the TS document contained the process of translations (Callon 1986) that had been required to reach the current status. In this way, it also provided its readers a means for legitimate peripheral participation (Lave 1991) in this particular community of practice, the negotiators of the sales.

To sum up, the technical specification is one of the key boundary objects between the customer and Valmet. Established tradition in the negotiators' community of practice means that the uses of the technical specification at this boundary are relatively well defined, and the inherent problems are understood, even though not simple or transparent. The TS changes constantly, and when it finally reaches closure during the early stages of the project, it also becomes secondary to designs and other documents that depict in more detail what is actually built. Between Valmet and the customer, the technical specification is initially (intentionally) weakly structured, to invite negotiations, but it becomes strongly structured, as it becomes fixed in the system of related boundary objects (Star 1989) within a community.

5. Conscripting Participants Within a Community of Practice

Within Valmet, the technical specification is a boundary object between the different units. Within the project team, however, the technical specification has a special role as one of the conscription devices (Henderson 1991, 1998b), enlisting and constraining participation in creating and maintaining it. In this, the technical specification is similar to a set of drawings for a turbine engine (Henderson 1991): its users must engage in inputting its elements and in revising them, if the specification is to serve its intended function for their particular purposes. Conscription devices are receptacles for knowledge created and adjusted through group interaction, aimed toward a common goal. Like all boundary objects, as they represent the group's negotiated ideas, they also structure how work is done in groups. We assumed that as a conscription device, the TS would not only support the coordination of the design and manufacture of this particular paper machine between various individuals and groups within Valmet and its subcontractors, but that it would also contribute to developing and strengthening their knowledge domain and practices, their perspective making. According to Boland and Tenkasi (p. 356), as a perspective strengthens, it "complexifies," that is, the detailed views of the machine implementation would start to emerge and take shape. However, with the TS, this did not take place, for reasons outlined next.

Even though the information in the technical specification was collected from the units that would use it again during the project, and even though the sales team consulted the units on changes during the negotiation process with the customer, the designers in the units still could feel that the connection between the information in the TS and the unit was lost or that involving the designers in the TS came too late in the project. Also, the machine concepts are likely to have changed while the sales negotiations took place. The PM engages in careful refeeding of the information back to the units when the project starts, so that the design teams can relate their detailed design guidelines to the TS. The TS alone is not sufficient to enlist the project team members to the project, but the input of the PM is vital.

> *The technical specification is the backbone for all we do. Feeding information to the project members is one of the most essential issues in the beginning....I usually hold these reading meetings, which ensure that the team members have received the documents, that they have read them, and that they have even understood them. Because very often the machine is different from the standard machine, I need to ensure that they have noticed the differences....In the beginning of the project, studying the technical specification takes up an enormous*

*amount of time, because one needs to learn it and read through it all.
But then, during the project, less and less. During the first weeks it is
very important, you almost study it by heart, so that you yourself have
internalized it and then you go through it with the guys. It is one of the
main means for us to get the message through to them. [PM of
Project G]*

When the project team members return to their units, to the unit project teams, they engage in perspective making within that community of practice. In this process, they relate the specifications to the paper machine concept and part standards in their design guidelines, listing out the differences. The head designer then discusses these differences with the PM, after which the unit project teams start making the designs. They may still use the TS, but finding relevant information in it requires practice, due to its size and structure.

*It comes to us only when the project is starting. There are usually
shortcomings. An item can perhaps be found there, but it is not
included in the price calculations. Then it will go free or we need to
invent a bit to get the money. It is like this also in the Project G. There
was a whole mechanism in one part missing in the price calculations,
even though that is included in the standard....We have to ask the PM
whether this and this is included and he then asks the sales department....Reading the technical specification is pretty much about
picking out information of what has been sold. It should be relatively
accurate because it is the starting point for design phase. Usually it
is according to the standards, but there can be a lot of extra sold and
we need to dig it out from there. [12]*

*The structure of the technical specification drives me crazy, it is so
comprehensive. Information from different units is just piled up in
there under separate headings, the information you are looking for
can be found scattered in several places, you need to leaf through the
thing. No page numbers in the table of contents, it is difficult to find
anything. [6]*

Meetings with the customer can result in changes to the TS, and the usual procedure for the PM is to inform those concerned quickly and directly. The information is then confirmed, for example, in meeting minutes, or internal orders (job number list changes). This also relates to negotiating the relationship of TS to other conscription devices. The PM of Project G gave primacy to meeting minutes and job numbers, due to the size of the project. The project team members could then trace changes by reading these and the related correspondence, but this was experienced by some as not coherent enough, and requiring too much attention to find the relevant changes among all the information. Also, as soon as the customer requirements are met and the interfaces of the part to other parts are specified, they become translated into set conditions for the part and the specification loses its importance for this part.

What the machine is, is then written into the operating and mainte-nance manuals. They are quite important for us. When they are done, nobody looks at the technical specification any more, it might have old information and cause errors. The information in the manuals is very detailed, but the technical specification is not. If the technical specification was to be updated all the time, it would end up being equally detailed. The actual machine...is different on the detail level, it has a different tailoring, even though on the surface it looks quite the same. [12]

To sum up, the technical specification acts as one of the conscription devices for the project members. The design teams clearly engage in mutual perspective taking with the TS, the design guidelines, and their sketches and drawings (Henderson 1991). However, in the company-level Valmet project team, this appears to be more problematic, due to the independence of the units (of whom the team members are only representatives), the planned nature of the project with boundaries spelled out, and the goal of sticking to the concept and standards. Only a minimum of information is exchanged, and that is usually related to exceptions. Thus the ideal situation, where "distinctive individual knowledge is exchanged, evaluated, and integrated with that of others in the organization" (Boland and Tenkasi 1995, p. 358) may not take place during the project (even though it may have taken place earlier, during R&D for the machine concept), and the "complexification" through perspective making may be hampered. The TS, for its part, does not support this, as it is a part of the contract and therefore its constant update and refinement is restricted. Therefore the TS is supported and gradually replaced by other conscription devices: internal orders and job number lists, designs drawings, database files, meeting minutes, correspondence, manuals, and the like.

Having a complete, up-to-date technical specification on paper or in a file, in one place, available to all, would thus be practical only in the beginning of the project. Also, as a set of documents, it is unwieldy and complex, and updating it would probably end in inconsistencies between different parts. Therefore, it is gradually left to the role of telling what was agreed in the beginning, with other documents taking over. When seen in this way, the necessary vagueness of technical specification as a conscription device loses its potential dangers within Valmet, and a "graceful departure" from it becomes a more focal issue.

6. Discussion

The main focus of our study was on establishing the value of the two concept pairs: boundary object and conscription device and perspective making and taking, in tackling the coherence-flexibility dilemma in designing computer support for collaboration. Both concept pairs have shown their fruitfulness on their own (Boland and Tenkasi 1995; Henderson 1991, 1998a, 1998b; Mambrey and Robinson 1997; Star 1993), but they have not been used together before. The two pairings that we studied with our case were the relationship of boundary objects to perspective taking and the relationship of conscription devices to perspective making. We started with two statements to focus our attention:

- When knowledge is made available to others, it becomes a boundary object, providing a basis for perspective taking, for communication that improves the ability of the community to take the knowledge of other communities into account.
- When participants in a community of practice engage in creating and maintaining a conscription device, they engage in perspective making, in strengthening the unique knowledge of their community.

6.1 Perspective Taking and Boundary Objects

With our case study, the first pairing informed us of several issues. First, when we looked at the technical specification as a boundary object between Valmet and its customer, it appeared flexible enough to invite the necessary negotiations between the parties, i.e., for perspective making in this community of practice, yet sufficiently strongly structured to act as the statement of commitments of a party, i.e., as material for perspective taking. Second, the negotiations with the standardized methods over a boundary object, the TS, gave each party an opportunity to give information for the perspective taking of the other party. That is, in this "perspective giving" the extent of perspective taking of one party was "allowed" by the other party. Third, when the history of the negotiations was explicitly spelled out in the TS, the perspective taking extended also to past negotiations, and allowed the TS as a boundary object tp span further than the negotiations at the moment. In terms of learning, the negotiations and the TS with its included history provided an important means for learning not only for negotiators, but also to related parties who followed them, in the form of legitimate peripheral participation (Lave and Wenger 1991). Finally, the TS was combined from many sub-documents and it was itself part of the contract. These documents could be studied as a system of boundary objects (Star 1989), to outline their connections, also in terms of further perspective taking.

The first pairing also brought attention to possible problems of increased support for perspective taking. Paper machinery delivery projects form a nexus of engineering competence in Valmet. If the information and knowledge that is necessarily involved in this is made visible to a larger audience than before and if thereby the processes become more transparent, it is likely to have consequences in the power relationships between the communities of practice (Zuboff 1988). The standardized methods (Star and Griesemer 1989) that are being used between Valmet and the customer, with shortcuts and leaks and playing with time, could be jeopardized by too much transparency. Examples of the negative consequences for Valmet could be loss of slack time that is used to give room for recovery of (unavoidable) errors, and loss of ability to smoothen the wrinkles in own performance prior to making it public. These kinds of losses could be detrimental for the whole company, limiting its flexibility and responsiveness.

Henderson tells of people who feared that small errors would have monumental consequences, and how these fears created an atmosphere of secrecy: "The fluidity and flexibility that is part of the loose structure of boundary objects was paralyzed by the fact that the whole system was computerized....The huge size and complexity of the interlocking systems intimidated people" (Henderson 1991, p. 464). If the TS was more complete and detailed, by the sheer amount of interrelated information, the TS also would be at risk to losing its informing capability across the Valmet-customer boundary.

Not only the flexibility of interpretations could be at risk, but it would also be possible that any contradictions, possibly existing as dormant in the Valmet-customer relationships, could unexpectedly surface.

To summarize, boundary objects and perspective taking pairing helped to draw attention to

*   the necessary flexibility and openness of the boundary object,
*   the need to limit communication with it at the same time,
*   the gradual, dialectical nature of the negotiation process,
*   the boundary object as informing the negotiation and as its result, and
*   how the boundary object was employed in perspective taking (and giving).

6.2 Perspective Making and Conscription Devices

The second pairing, conscription device and perspective making, also informed us of several issues. First, perspective making took place with the TS only early in the project, with the active intervention of the project manager, due to the nature of the Valmet project team. It met as a whole only in the beginning of the project and, after that, the communication was mainly between the PM and each individual project team member. The Valmet project team was only a short-term virtual construct,[1] not a long-living "community of knowing" (Boland and Tenkasi 1995). The permanent communities of practice for the team members were their respective design groups.

Second, because the creation of the TS was unidirectional, and it could only be changed via negotiations with the customer, other documents, kept fully within Valmet, gradually replaced it as the description of the paper machine to be. These other documents, therefore, become conscription devices, more focal to perspective making.

Third, due to the relatively general level of description, the TS was not very useful to the design teams, after the layout of the machine was fixed. After that, the current design guidelines for the type of machine became the guidelines to follow, with the possible risk that the current machine concept would be more influential in design than the machine ordered. However, this risk was actually not a risk, since Valmet openly professes to supply the latest technology, whether the customer knew to ask for it or not. The preference to design guidelines was further enhanced by the cumbersome format of the TS: finding relevant information in the TS was difficult and required considerable practice.

Fourth, information about the changes to the TS were distributed by several means. Keeping track on all relevant changes required active reading, and even then some may have passed unnoticed. This can be taken to indicate that, after all, a comprehensive, up-to-date, detailed description of the machine, accessible to project team members, could

---

[1]The problem of the loose connection between members of the Valmet project team has been addressed already with the Tasman application (Karsten et al. 1997), designed to hold most of the shared documents in the project; that is, the virtual nature of the project team has been acknowledged and it has guided the kind of support mechanisms that have been built. Our case, Project G, was one of the first project teams to experiment with Tasman use Valmet-wide, holding meeting minutes and monthly reports there. A wider coverage and use were still only discussed, to be addressed in later projects.

be the kind of conscription device that would aid perspective making within Valmet. However, this comes with risks attached, according to Henderson, who has drawn attention to boundary objects that could be too comprehensive and accessible. In her view, the difficulties the employees encountered in interacting with a repository reflected the difficulties the departments had in interacting with one another. If there were problems between Valmet units, they could become more visible with this kind of machine description.

The plans in Valmet seemed to approach this issue via the ERP system, which would tie together not only the specifics of the machine, but also connect it to more formal data such as in budgeting, resource planning, ordering, and scheduling. The problem of extensive coherence across the whole of Valmet, and thereby decreased flexibility for units, however, directs the attention to less controlled alternatives, such as document databases with (hyper)links to the relevant, more formal data, possibly in the ERP system. These would shift the balance to the other direction, as document-based, semi-formal and loosely structured alternatives could result in loss of controllability and manageability of the information.

In summary, conscription devices and perspective making pairing drew attention to the fact

- that loss of ownership, limited maintainablity, decreasing relative importance, and wieldy structure narrowed the usability of the TS as a conscription device in perspective making;
- that perspective making was seen to take place only in "real" teams, such as design teams, but only with special effort in "virtual" teams such as the Valmet project teams;
- that a "functional" conscription device could have the potential of supporting perspective making but also dangers; and
- that in considering the technical solution, how coherence and flexibility are balanced has consequences in perspective making, and vice versa.

7. Conclusion

The approach taken in this study was to look at one set of documentation, the technical specification of a paper machine, in crossing the boundary between Valmet and the customer, and in conscripting participation in the Valmet project team. On the boundary, it proved to be flexible enough as a boundary object, and diffuse enough to invite the necessary negotiations over it, with the resulting perspective taking. However, as a conscripting device within Valmet, it had several shortcomings, especially in relation to its limited changeability, to its relatively diffuse relationships to other conscripting devices, and to the practical difficulties in using it.

In parallel to Brown and Duguid's observations, in our case, the information was moving within Valmet in a rather sticky way, but much more fluently between Valmet and the customer. Brown and Duguid's explanation was that knowledge is continuously embedded in practice and thus circulates easily within a community of practice. Between the negotiating parties of Valmet and the customer, the long established practices of sales, contract negotiating, and cooperation during the project may have made them *de facto* communities of practice. There are not many paper producers with which Valmet has not had projects, and even when the individual persons have changed, the practices have remained to a great extent the same. Within Valmet, however, the units are geographically dispersed, and after the organizational changes of the past years, they are prone to establish their own practices in supplying parts or services for the paper machine projects, not necessarily in alignment with those of other units.

In terms of designing computer support for collaboration, our approach drew attention to several issues. First, the nature of boundary objects and conscription devices appeared to be open and modifiable in perspective taking and making. As soon as any deterrents were put in place to make changes difficult, the objects seemed to lose their function. Second, the dialectic between the object and the process seemed to work in both directions: the conscription device was the means of perspective making and perspective making resulted in changing the conscription device; and the same appeared also for boundary object and perspective taking. A concern related to this is how to increase action visibility, while at the same time acknowledging the need to bound the resulting transparency.

Our study thus indicates that balancing of flexibility and coherence is bound to be an active, situated process. Our study gave no directions as to how this dilemma could be solved in practice; it only drew attention to the dialectic nature of boundary objects and conscription devices in perspective making and taking. Thus our view of design process correlates with that of Bowers (1991), who sees systems development to be relational, reflexive, critical, and practical; proceeding from locally constructed and modifiable solutions, to shared work practices spread via interlinked and intertwined communities of practice, and in this way balancing coherence and flexibility.

To conclude, the key dilemmas we have identified are, how to ensure the availability of correct and sufficient information when needed, but at the same time provide enough background to assess the information; and how to increase action visibility, while at the same time acknowledging the need to bound the resulting transparency. The problem is also to understand the contingencies that affect the level of coherence and flexibility needed. An area for further work would be viewing these issues from the information system implementation perspective, with a consideration for technological resources and infrastructures.

Acknowledgments

We would like to thank Valmet and all our informants there for the free and generous access to their activities. Special thanks go to Mr. Jorma Hujala, head of the project department during this study, and to the project manager of Project G, for comments and clarifications on the earlier versions of this paper. The reviewers of the conference have also given valuable, much appreciated, feedback.

References

Altheide, D. L., and Johnson, J. M. "Criteria for Assessing Interpretive Validity in Qualitative Research," in *Handbook of Qualitative Research*, N. K. Denzin and Y. S. Lincoln (ed.). London: Sage, 1994, pp. 485-499.

Boland, R. J., Jr., and Tenkasi, R. V. "Perspective Making and Perspective Taking in Communities of Knowing," *Organization Science* (6:4), 1995, pp. 350-372.

Bowers, J. M. "The Janus Faces of Design: Some Critical Questions for CSCW," in *Studies in Computer Supported Cooperative Work: Theory, Practice and Design*, J. M. Bowers and S. D. Benford (eds.). Amsterdam: North-Holland, 1991, pp. 333-350.

Brown, J. S., and Duguid, P. "Organizing Knowledge," *California Management Review* (40:3), 1998, pp. 90-111.

Callon, M. "Some Elements of a Sociology of Translation: Domestication of the Scallops and the Fishermen of St. Brieuc Bay," in *Power, Action and Belief*, J. Law (ed.). London: Routledge and Kegan Paul, 1986, pp. 196-233.

Henderson, K. "Flexible Sketches and Inflexible Data Bases: Visual Communication, Conscription Devices, and Boundary Objects in Design Engineering," *Science, Technology and Human Values* (16), 1991, pp. 448-473.

Henderson, K. *On Line and on Paper: Visual Representations, Visual Culture and Computer Graphics in Design Engineering.* Cambridge, MA: MIT Press, 1998a.

Henderson, K. "The Role of Material Objects in the Design Process: A Comparison of Two Design Cultures and How They Contend with Automation," *Science, Technology, and Human Values* (23:2), 1998b, pp. 139-174.

Karsten, H.; Lyytinen, K.; Heilala, V.; and Tynys, J. "The Impact of User Support in Successful Groupware Implementation: Case Tasman to Support Paper Machinery Delivery," a paper delivered at the *ECIS'97, The Fifth European Conference on Information System*, Cork, Ireland June 19-21, 1997.

Lave, J. "Situated Learning in Communities of Practice," in *Perspectives on Socially Shared Cognition*, L. R. Resnick, J. M. Levine, and S. D. Teasley (eds.). Washington, DC: American Psychological Association, 1991, pp. 63-82.

Lave, J., and Wenger, E. *Situated Learning: Legitimate Peripheral Participation.* New York: Cambridge University Press, 1991.

Mambrey, P., and Robinson, M. "Understanding the Role of Documents in a Hierarchical Flow of Work," in *GROUP'97: International ACM SIGGROUP Conference on Supporting Group Work: The Integration Challenge*, S. C. Hayne and W. Printz (eds.). Phoenix, AZ: ACM, 1997, pp. 119-127.

Mathiassen, L., and Stage, J. "The Principle of Limited Reduction in Software Design," *Information Technology and People* (6:2/3), 1992, pp. 171-186.

Nandhakumar, J., and Jones, M. R. "Too close for Comfort? Distance and Engagement in Interpretive Information Systems Research," *Information Systems Journal* (7), 1997, pp. 109-131.

Ngwenyama, O. "Groupware, Social action and Emergent Organizations: On the Process Dynamics of Computer Mediated Distributed Work," *Accounting, Management and Information Technology* (8:4), 1998, pp. 123-143.

Robinson, M. "Computer Supported Cooperative Work: Cases and Concepts," in *Readings in Groupware and Computer Supported Cooperative Work*, R. Baecker (ed.). Palo Alto, CA: Morgan Kaufman, 1991.

Star, S. L. "The Structure of Ill-structured Solutions: Boundary Objects and Heterogeneous Distributed Problem Solving," in *Distributed Artificial Intelligence, Volume 2*, M. Huhns and L. Gasser (eds.). London: Pitman, 1989, pp. 37-54.

Star, S. L. "Cooperation Without Consensus in Scientific Problem Solving: Dynamics of Closure in Open Systems," in *CSCW: Cooperation or Conflict?*, S. M. Easterbrook (ed.). London: Springer Verlag, 1993, pp. 93-106.

Star, S. L., and Griesemer, R. J. "Institutional Ecology, 'Translations', and Boundary Objects: Amateurs and Professionals in Berkeley's Museum of Vertebrate Zoology, 1907-39," *Social Studies of Science* (19), 1989, pp. 384-420.

Taylor, S. J., and Bogdan, R. *Introduction to Qualitative Research Methods: A guidebook and Resource*, 3rd ed. New York: Wiley, 1998.

Wastell, D. G. "The Social Dynamics of System Development: Conflict, Change and Organizational Politics," in *CSCW: Cooperation or Conflict?*, S. Easterbrook (ed.). London: Springer-Verlag, 1993, pp. 69-92.

Yin, R. *Case Study Research: Design and Methods*. Newbury Park, CA: Sage, 1989.

Zuboff, S. *In the Age of the Smart Machine*. New York: Basic Books, 1988.

About the Authors

Helena Karsten has done research and teaching at the University of Jyväskylä in Finland since 1993. Prior to that she worked at the Technical Research Centre of Finland for many years. Her research area is collaborative information technologies and organizational change. She is a research fellow in the three-year Globe project, studying management of paper machinery delivery projects, seeking to understand the role of IT in global virtualiation of work. Helena's e-mail address is eija@jytko.jyu.fi

Kalle Lyytinen is Professor of Information Systems at the University of Jyväskylä, Department of Computer Science and Information Systems, since 1987. He is also the Dean of the new Faculty of Information Technologies in Jyväskylä. Kalle is the head of the Globe project. He is the former chair of IFIP 8.2 and a prolific researcher who has published over 80 research articles and written or edited six books. He is a current member of several editorial boards in leading IS journals and currently serves as a senior editor of *MIS Quarterly*. Kalle's e-mail address is kalle@jytko.jyu.fi

Markku Hurskainen received his Masters in Economics and Business Administration degree in 1998 at University of Jyväskylä, majoring in Information Systems. Markku has worked since then in Solution Garden Ltd., a consulting company, specializing in knowledge management systems for process industry. Markku's e-mail address is markku.hurskainen@solutiongarden.fi.

Timo Koskelainen received his Masters in Economics and Business Administration degree in 1998 at University of Jyväskylä, majoring in Information Systems. Timo has worked since 1997 in Valmet as a development engineer in the Project Department, on issues of information technology and project management development. Timo's e-mail address is Timo.Koskelainen@valmet.com.

16 THE ROLE OF INFORMATION TECHNOLOGY IN THE LEARNING OF KNOWLEDGE WORK

Valerie Spitler
New York University
U.S.A.

Michael Gallivan
Georgia State University
U.S.A.

Abstract

Knowledge work is increasingly important in post-capitalist society (Drucker 1993, 1995) and is associated with new organizational forms and ways of working (Lucas 1996). These include flatter, less hierarchical organizational structures; more fluid job definitions and reporting structures; more competitive and faster paced work environments; and an increased reliance on information technology (IT) to perform work (Ruhleder, Jordan and Elmes 1996). Initiating knowledge workers into such firms and working environments and ensuring their continued performance and productivity will be critical to firms and may require new management practices. The purpose of the present research is to develop a theoretical framework which explores and describes how knowledge workers learn their jobs and the role that using IT plays in this process. Founded on the assumptions that learning and problem-solving are critical to knowledge work, and are socially constructed, situated in practice, and context-specific, the research presented here is part of an on-going interpretive case study based on theoretical underpinnings derived from the theory of legitimate peripheral participation (Lave and Wenger 1991). To develop this framework, one of the authors has gained access to a global strategic management consulting firm with offices in New York

City where she is using ethnographic methods of interviewing and participant observation. This paper presents early results of the study, focusing specifically on how entry-level consultants, analysts and associates, learn and perform their jobs, and the role that using IT plays therein. The research is expected to have implications for training, mentoring and incentive policies for organizations operating in the IT-based, knowledge economy.

Keywords: Knowledge work, information technology, situated learning, interpretive case study, IS usage, community of practice.

> *"And even if outnumbered by other groups, knowledge workers will be the group that gives the emerging knowledge society its character, its leadership, its social profile."*
> *Drucker 1995, p.233*

> *"The knowledge [worker] may need a machine, whether it be a computer, an ultrasound analyzer, or a radio telescope. But neither the computer nor the ultrasound analyzer nor the telescope tells the knowledge [worker] what to do, let alone how to do it. Without this knowledge, which is the property of the employee, the machine is unproductive."*
> *Drucker 1993, pp. 64-65*

1. Introduction

Knowledge work is increasingly important in the post-capitalist (Drucker 1993, 1995), post-industrial (Bell 1973), information age society. Not only do knowledge workers make up an increasing percentage of our labor force, but they also embrace an important leadership role in society (Drucker 1995). Organizations are striving to understand and manage knowledge work (Davenport, Jarvenpaa and Beers 1996; Zand 1997) and the role that information technology (IT) plays in performing it (Fisher and Fisher 1998; Mankin, Cohen and Bikson 1996; Orlikowski 1996). Their attempts have illuminated the complexity of knowledge work and workers' interaction with technology (Orlikowski 1992), and have highlighted the social-cognitive nature of the work (Barley 1990; Tyre and Orlikowski 1994).

Knowledge work is associated with new organizational forms and ways of working (Lucas 1996). These include flatter, less hierarchical organizational structures; more fluid job definitions and reporting structures; more competitive and faster paced work environments; and an increased reliance on IT to perform work (Ruhleder, Jordan and Elmes 1996). Initiating knowledge workers into such firms and working environments and ensuring their continued performance and productivity will be critical to firms and may require new management practices.

The purpose of the present research is to develop a theoretical framework which explores and describes how knowledge workers learn their jobs and the role that using

IT plays in this process. Founded on the assumptions that learning and problem-solving are critical to knowledge work, and are socially constructed, situated in practice, and context-specific, the research presented here is part of an on-going interpretive case study based on theoretical underpinnings derived from the theory of legitimate peripheral participation (Lave and Wenger 1991). To develop this framework, one of the authors has gained access to a global strategic management consulting firm with offices in New York City where she is using ethnographic methods of interviewing and participant observation.

This paper presents early results of the study, focusing specifically on how entry-level consultants, analysts and associates, learn and perform their jobs, and the role that using IT plays therein.

The research is expected to have implications for training, mentoring and incentive policies for organizations operating in the IT-based, knowledge economy.

2. Assumptions Underlying the Research

The present research deviates from the dominant view of IT use as occurring with a single system or set of functions, by a single, isolated individual, and to a greater or lesser degree. Such a view is appropriate for studying the interaction of a single individual with a single system such as in cognitive science or human factor studies (Kling and Scacchi 1982), but is inappropriate for understanding how workers interact with IT in an organizational setting. Thus a social-cognitive perspective is proposed here in order to understand the relation between knowledge workers and IT. This perspective relies on the notion of *communities of practice* (Brown and Duguid 1991; Lave and Wenger 1991), which "imply participation in an activity system about which participants share understandings concerning what they are doing and what that means in their lives and for their communities" (Lave and Wenger 1991, p. 98). Knowledge workers do not use IT in an isolated manner, but rather operate within communities of practice, whether it be lawyers using a document management system within the law practice to prepare a client case, stock brokers and sales assistants using multiple database and analysis systems within an investment management firm to make invest-ment recommendations, or management consultants using a variety of desktop informa-tion technologies working in teams to understand and solve client problems.

This research proposes that, at the same time that IT use occurs within a community of practice, it is also more complex than typically viewed in the IS literature. IT use among knowledge workers need not always be limited to a single system or set of func-tions, even when knowledge workers are solving a single problem. Further, given their higher levels of education and training (Drucker 1995) and the unstructured and creative nature of knowledge work, knowledge workers are more likely to manipulate informa-tion technology to meet the demands of their work. Manipulation may include not only changing options and parameters of systems (Devin 1994), but also (1) modeling relations in the data, (2) combining information and functionality from multiple sources, (3) revising parameters based on reasonableness of data, etc. Thus, use of IT is ex-panded to include not only a relatively passive activity of using an information system where data, features and functionality are relatively fixed and well designed to support

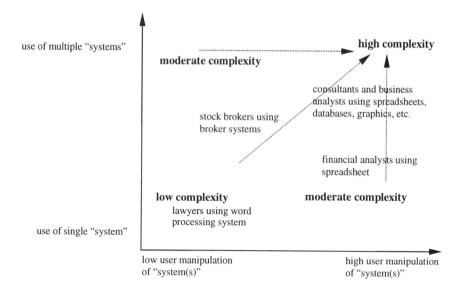

**Figure 1. Examples of Knowledge Workers and Their Use of IT
(working within community of practice)**

specific structured work (Turner 1984), but also as an engaging activity where the workers, in the extreme, may determine which particular technology to use, which data to include, which features and functions to use and what parameters and options to change. The view of IT use proposed in this research is depicted in Figure 1.

The bottom axis represents the degree of manipulation users make to their information technologies, while the left axis represents the number of systems or functionalities they use. Much of early IS research investigated workers using a transaction processing system; workers using such a system are limited to using a single system and make minimal manipulations to the system. This type of use falls in the lower left quadrant of the diagram and seems to represent the assumption underlying much of current IS research. Knowledge workers' use of IT today is generally more complex than that of these early users. The trend is for their use of IT to become more complex, for them to use multiple information technologies and to perform a high degree of manipulation. This highly complex use is represented in the upper right quadrant of the diagram and represents the type of knowledge workers who are the subject of the present research.

This research proposes that our understanding of how IT is used to support knowledge work can be enhanced if we adopt a characterization of *IT use* as
- having a high intellectual (cognitive) component,
- occurring within a community of practice,
- being both a medium and an outcome of human action, enabling and constraining it,
- varying in its complexity in terms of the number of "systems" (i.e., functionalities) used and the degree of manipulation of those "systems."

While the first three characterizations have been discussed explicitly in the literature, the last is an assumption unique to the present research.

Viewing the use of IT as a social-cognitive activity underlies and guides the present research. This perspective should provide valuable insights over and above those achieved by many prior studies. It allows examination of workers actively engaged in using IT in their natural environments. The intention of the research is to reveal and explore fundamentally how knowledge workers use IT for performing their work, while avoiding the adoption of simplistic and limiting assumptions.

3. Information Technology Use in the IS Literature

There is a growing concern within organizational, operations management and IS literature about the role that IT plays in performing knowledge work (Davenport, Jarvenpaa and Beers 1996; Drucker 1995; Laudon and Starbuck 1994). This type of work and the role that IT plays in performing it, however, has not been adequately explored. The results which can be gleaned come from two types of IS research. The first are positivist studies which attempt to predict workers' use of IT (see, for example, De Lone and McLean 1992), usually as part of the implementation research. The second type of IS research views workers' use of IT as integrated into their jobs, often by the workers themselves. This integrated perspective is represented in the work of researchers such as DeLong (1997), Orlikowski (1992, 1996), and Schultze (1999) takes an interpretive perspective, and is often concerned with organizational changes associated with using IT. Given the relatively recent expansion of knowledge work in our economy, as well as the variety of IT to support it, and the few interpretive studies on knowledge workers, it is not surprising that few research models or results are forthcoming.

One theory in particular, *legitimate peripheral participation* (Lave and Wenger 1991), has emerged in the cognitive anthropology literature and deals with the situated nature of cognition, problem-solving, and learning, and thus seems valuable for the study of the use of IT for performing knowledge work in organizations. This theory, which helped explain differential learning among two groups of workers using desktop IT (George, Iacono and Kling 1995), is expanded and developed here, as it applies to knowledge workers, in order to guide the present research. The intent of the present research is to understand the role that the use of IT plays in knowledge workers' learning and performing their jobs.

4. Theoretical Framework: Legitimate Peripheral Participation

Both scholars and practitioners have discussed the nature of knowledge work. This work has been described variously as non-routine and creative (Davenport, Jarvenpaa and Beers 1996), involving human information processing (Davis 1991) or cognitive skills (Sulek and Marucheck 1994), and requiring "continuous learning" (Drucker 1995, p. 226). This type of work contrasts with more traditional forms of work where jobs are designed with well-defined, fixed, specific and routine tasks. Typical tasks of knowledge work are planning, decision making, problem solving and communicating, which are ill-defined and ill-structured. These tasks are generally performed using desktop information technology.

No comprehensive theories of knowledge work or the role that IT plays in performing it exist, but one situated learning theory, which may be adapted for this purpose, outlines how cognition, learning and problem solving occur in situ, rather than in the laboratory or in the classroom (Lave 1988; Lave and Wenger 1991). This theory of situated learning, legitimate peripheral participation (LPP), attempts to explain thinking and learning as a social and practical matter rather than as an individual and academic phenomenon. Given that knowledge work has a large cognitive component, involves continuous learning and occurs in an organizational setting, this theory provides a useful starting point for understanding knowledge work and the role that IT plays in performing it.

LPP has been developed from studying other types of workers and learners such as butchers, naval quartermasters and supermarket shoppers (Lave 1988; Lave and Wenger 1991; Rogoff and Lave 1984), and it includes a role for technology in general, but not for information technology specifically. Except for the study by George, Iacono and Kling, the theory has not been applied to knowledge workers. In that study, community of practice was evident in a group of knowledge workers (financial planners) and was given as the reason a group of knowledge workers used IT to serve them effectively while a group of clerical workers did not. The finer details of the theory have not been developed for knowledge workers, and the development of a role for IT in the theory has not been forthcoming.

LPP is characterized as participation in the social world and has the following main features: learning within a community of practice; learning involving the whole person and the construction of identities; a diverse field of actors including newcomers and old-timers, where newcomers learn from old-timers and peers but where old-timers also continue to learn; and learning in practice where learning takes place in productive activities.

> [LPP] provides a way to speak about the relations between newcomers and old-timers, and about activities, identities, artifacts, and communities of knowledge and practice. [LPP] concerns the process by which newcomers become part of a community of practice....[LPP] is proposed as a descriptor of engagement in social practice that entails learning as an integral constituent. [Lave and Wenger 1991, pp. 29, 35]

The choice of terms to describe this theory is revealing. "Participation" connotes the idea that learning is not the traditional notion of transfer of knowledge or skill from one person to another, or from one situation to another within the same person; rather it takes on a larger meaning of participation by the "learner" within the social world, within a community of practice. "Legitimacy" suggests the "learner's" right to participate, and the forms that that participation may take. Apprentices are generally legitimate members of their communities of practice. However, gaining legitimacy may be a problem for newcomers in some situations. For example, hazing, is one form of denying legitimacy; during hazing learning is inhibited. Only once they have survived hazing and gained legitimacy do newcomers occupy a place of legitimate participation, but the forms of their legitimacy of participation may vary. This latter concept is better understood when taken together with the notion of "peripherality." Peripheral participation may be contrasted with *full participation,* that to which peripheral participation leads. Peripherality suggests that there is no center of a community of practice and that "there are multiple, varied, more- or less-engaged and -inclusive ways of being located in the fields of participation defined by a community" (Lave and Wenger 1991, p. 36) Peripheral participation is a dynamic concept and implies that the activities of "learners" are connected and relevant to those of the community. The three terms, legitimate peripheral participation, are meant to represent a concept to be taken as a whole.

LPP includes a role for technology, but not for IT specifically:

> Becoming a full participant certainly includes engaging with the technologies of everyday practice, as well as participating in the social relations, production processes, and other activities of communities of practice. The understanding to be gained from engagement with technology can be extremely varied depending on the form of participation enabled by its use. [Lave and Wenger 1991, pp. 102-103]

In the current research, IT is given a primary role, since it is seen as critical for performing knowledge work. Some organizations engage in such knowledge work regularly and have taken the lead in adopting various learning strategies using IT. The current, on-going research addresses three specific aspects of IT in knowledge work: (1) learning how to appropriate IT in knowledge work, (2) knowledge workers' identity with respect to using IT, and (3) the transparency of IT in knowledge work. Although LPP relates all three aspects, only the first is covered in this paper.

5. Research Approach, Design and Methodology

This research is based on the assumption that our knowledge of reality is socially constructed. Thus, the approach is to gain understanding of phenomena through the meanings people assign to them by studying one organization in-depth, rather than to use a survey technique across multiple organizations.

The study can be described as an in-depth, interpretive case study (Klein and Myers 1999) with embedded units of analysis (Yin 1988), but also relies on ethnographic methods such as participant observation (Agar 1996). While the design of the study is bounded by the topic under investigation (Miles and Huberman 1994), it is also a flexible design, molding to the specific organization, its design, work practices and culture (Janesick 1998).

The first author gained initial access to the firm in early October 1998. With the exception of not being able to attend client meetings or going to client sites, and the requirement to be unobtrusive, her access is relatively unrestricted, limited to her own initiative, imagination and time constraints. The primary data collection techniques to date have been formal interviews (since consultants travel frequently and researcher access to client sites and meetings is not permitted) and participating in a week-long, off-site training session for new consultants. Secondary techniques include observations, collection and interpretation of firm documents, use of the firm-issued laptop computer, inclusion on voice and electronic mail distribution lists, and participation in some of the firm's social activities. Consistent with ethnographic methods, the researcher also keeps a detailed set of field notes, recording how she spends her time at the site, her observations and thoughts, and issues for ongoing inquiry. To date, she has conducted 30 interviews, as follows:

Position	Number of People Interviewed
New analyst (less than one year with the firm)	9
New associate (less than one year with the firm)	5
Second-year analyst (between one and two years with the firm)	9
Associate (promoted from analyst, more than two years with the firm)	1
Staff partners	2
Support staff	4
Total	30

During each interview, the researcher reminds the participant of the confidentiality of the interview and explains how anonymity will be maintained. Interviews are based on the conversational partners technique (Rubin and Rubin 1995) and are semi-structured. They last 30 to 90 minutes and are tape recorded and transcribed. Research participants have the opportunity to offer corrections or modifications to transcripts. This process encourages participants to speak freely during the interview.

Data collection and analysis strategies are determined by following principles proposed by Klein and Myers (1999) and Strauss and Corbin (1990). Thus data collection and analysis require giving consideration to not only the views and behaviors of individuals in the firm under study, but also to the context in which these individuals operate (fundamental principle of the hermeneutic circle, Klein and Myers 1999). Using these principles, the researcher compares findings across different respondents, documents and observations, within the context of the consulting profession and the firm under study, and uses the theory of LPP as a guide. Data collection and analysis are intertwined; analysis occurs as the data is collected and leads to further data collection and analysis. The process itself is at least partially emergent (Myers 1999). Specific analysis techniques (Miles and Huberman 1994; Strauss and Corbin 1990) include (1) writing theoretical memos, (2) drawing concept maps and (3) coding interview transcripts and documents to identify concepts, categories and their properties.

The researcher guards against bias in several ways. First, the researcher seeks multiple perspectives from participants with different backgrounds, levels of experience, and job content (principle of multiple interpretations, Klein and Myers 1999). Second, by collecting data through multiple methods—interviews, observations and firm documents, she validates many aspects of the data through triangulation (Gallivan 1997) and demonstrates "sensitivity to possible biases and systematic distortions in the narratives collected from the participants" (the principle of suspicion, Klein and Myers 1999). Third, the researcher engages regularly in the comparison of findings with theoretical preconceptions (dialogical reasoning, Klein and Myers 1999), and documents the comparisons with theoretical memos.

The remainder of this paper provides some preliminary data and tentative conclusions from the research. Section 6 provides the context in which this group of knowledge workers operate by describing their firm, their work and their technology environment. Section 7 provides details about the specific work of new consultants, analysts and associates, as well as the training and other mechanisms by which they prepare for this work. Section 8 discusses the research in terms of particular aspects of the theory of LPP, while the last section considers how these preliminary findings relate to theories of IT use, acceptance and training in the IS literature.

6. Arris Strategic Management Consulting[1]

6.1 Industry and Firm

Strategic management consulting solves the strategic problems of companies, and is a profession based on intellectual capital. The major strategic management consulting firms are McKinsey, Booz Allen and Hamilton, Bain and Co. and Boston Consulting Group, whose clients are Fortune 500 and Fortune 1000 companies. Strategic management consulting is a high prestige, glamorous business where the work is intense and involves long hours and frequent travel. The pinnacle of a career in strategic management consulting is to become a partner in the firm. Many consultants, however, leave the business long before this stage to pursue other careers or interests. At Arris, in spite of the non-hierarchy, there is a strong distinction between partners and the consulting staff.

Like McKinsey and others, Arris is a global strategic management consulting firm with offices and clients throughout the world. Although Arris' origins go back several decades, the firm is relatively young, formed by mergers and acquisitions of smaller firms over the years. It is young in other ways as well: a large portion of its workforce has been with the firm for only two or three years; further, many of these people are in their twenties.

Each year Arris hires two types of entry-level consulting professionals: analysts and associates. Analysts are recent college graduates (with Bachelor of Arts or Bachelor of Science degrees) from elite institutions around the world. In the U.S., analysts are typically hired from the Ivy League institutions, as well as from other elite schools. Associates may be either recent M.B.A. graduates of top business schools or employees that have been promoted from the analyst position, either with or without an M.B.A.

---

[1]"Arris" is a pseudonym. The primary author was required to sign a confidentiality agreement with the firm, promising to disguise its identity. Therefore, some terminology and facts deemed insignificant for the analysis have been modified to disguise the identity of the firm and its members.

6.2 The Work of Arris's Strategic Management
Consultants: Client Projects

The main form of work for a strategic management consultant is a client project. The on-site researcher had the opportunity to participate in a simulated client project during a one-week training session. She was a member of an 11-member team of new hires (analysts and associates) that was headed by a junior partner. The team's mission, as put forth by the (mock) CEO of a major quick service food conglomerate, Binstar, was to propose a strategy to make Binstar's earnings grow by 10% within five years. Over the course of the next four days, her team analyzed the quick service food industry, including past and potential future trends in the industry and the competitive environment in the industry, brainstormed about possible ideas to explore further, interviewed (mock) Binstar managers and quick service food industry experts; analyzed data from a 2,000-person survey, created pro-forma financial statements for three of the most promising ideas, and prepared and presented a 50 page presentation to the CEO of Binstar (role-played by one of the Arris partners). The team leader often divided up the work for individuals or teams of two or three consultants to work in parallel on different aspects of an analysis. The team was provided with information packs progressively throughout the week which contained, for example, industry reports, Wall Street analysts' reports, Binstar's and other firm's annual reports, and a SAS cross-tabulation report for a 2,000-subject, 50-variable survey. Each consultant had been instructed to bring his laptop computer to the training (see section 6.3). A large portion of the work done during the week, especially economic and market analyses and presentation of ideas and results, was facilitated by consultants' use of their laptop computers. The researcher was not equipped with a laptop computer and as a result felt handicapped in participating in the work (Field Notes: 1-8-99).

The Binstar project represents the type and intensity of work in strategic management consulting. Client projects take a variety of forms and are typically staffed with a team of consultants with varying experience. The average length of a project is three to four months and deadlines with "deliverables" constantly loom over the team. Unlike the Binstar simulated project, where the team was composed of new hires, a typical project team includes a partner, a junior partner or senior associate, two or three associates and two or three analysts. In most cases, the roles of the different team members are as follows. The partner sells the project to the client and maintains the relationship with the client to set and maintain expectations. The junior partner or senior associate on the project acts as the team and project manager, assuring that work promised to the client meets expectations and is completed on time. This team manager also acts as a buffer between the partner and the consulting staff (i.e., associates and analysts) (Interview Transcript). Associates and analysts conduct research, perform market and economic analyses, and prepare written presentations.

To support this work, the New York office has a research library staffed with two full-time librarians, a production department to assist in producing client presentations, and an information systems department staffed with two full-time information systems technicians who service computers, printers and the local network. New York-based consultants are not limited to the support of New York-based resources. In particular, a marketing sciences group operates out of another office and assists with projects which require complex statistical or other analyses. Further, one consultant described how she uses the time difference of geographically dispersed offices to keep production (in the different offices) working around the clock on a time-critical document (Interview Transcript).

6.3 Information Technology Environment at Arris

Although several consultants have indicated that Arris is behind its competitors in terms of the information technology environment, some have been quick to point out that the environment has improved over the years. Each consultant is provided with a firm-issued laptop computer, which has been pre-configured by the IS department. A typical configuration includes Windows 95 operating system, Microsoft Office 97 (Excel, Powerpoint, Word, Access), Novell network access to Arris local and remote servers and printers, Microsoft Outlook 98 (e-mail), Microsoft Explorer for access to the Internet and to Arris intranet and knowledge management system, and pre-configured spreadsheets for tracking time and expenses. Powerpoint has been customized with the addition of over 100 custom templates, which are available via a special pull-down menu from within the Powerpoint application. Consultants sometimes use more powerful computers and specialized software such as SAS, SPSS or Knowledge Seeker.

Using IT is a natural part of consultants' work (but not necessarily of the partners' work), so much so that some consultants take it for granted. At the Binstar training, one of the new associates, knowing the general area of this study, commented "It's unfortunate for you that we didn't really use any IT during the project simulation." When the researcher referred to the laptop computers, the associate said, "Oh, I didn't think of that because we use it all the time" (Field Notes 1-8-99). Others realize that using IT is a necessary condition of their working lives. One of the new analysts expressed this view:

> I think that the more that information technology gets ingrained in what we do, when someone comes in without a capability, [he's] behind to a degree, so [he's] forced to learn it quicker just because everyone around him is using those programs. So, if five years ago we weren't using computers, that means now, if your interest wasn't in computers or IT, you have to learn that because that's a skill you have to have....[M]ost of the work we do in some way involves information technology, either the analyses or manipulating data. That goes on in Excel or sometimes statistical packages. [Interview transcript]

Arris is a firm where IT is integrated into consultants' lives. (For partners, it may be a different story, which we shall tell in future research!)

Using IT is a complex activity for consultants. They regularly work with a variety of information technology, including spreadsheets, databases, statistical software, presentation software, electronic mail, file transfer and download functions, and electronic searching software. Further, they interact with the information technology by building models, manipulating data, preparing charts, etc. Their interaction goes beyond data input and reading output to include, for example, searching for functions and writing them into formulas in Excel and statistical packages, writing and testing queries in Access, and manipulating fonts, objects, text and diagrams in Powerpoint. These knowledge workers clearly belong in the upper right quadrant of Figure 1.

7. First Job as a Strategic Management Consultant

7.1 New Analysts and Associates: Recruiting, Hiring, the Job

Recruiting is serious business since Arris runs on intellectual capital. As one partner expressed it: "It's a people-oriented business....We do not make products. We sell the intellectual capital of the firm" (Interview Transcript) For candidates, it is serious because a job at Arris provides lots of opportunity, not just while at Arris, but post-Arris, as well. One new analyst expressed his reasoning like this:

> I'm not really sure what I want to do with my life so this is probably
> something that will definitely be good for a little while, maybe for a
> long while, and won't hurt my chances in anything else if I decide
> there's something else I want to do. [Interview Transcript]

Strategic management consulting is a high burn-out business, and high turnover is expected. A partner explained it this way: "Consulting years are like dog years....You spend one in consulting; it's like seven in any other business" (Interview Transcript). Thus, each year Arris hires a new crop of young, fresh-faced, highly intelligent, well-educated, energetic, and sociable analysts and associates. Firm-wide, Arris hired approximately 150 new consultants in 1998.

Getting a job as a strategic management consultant is an achievement and a performance. The recruiting season for new analysts starts in the Fall, when potential candidates check out Arris' Web site, or more likely talk to a friend who is already at Arris. Candidates submit resumes (hoping to get past the initial screening), pass on-campus interviews, and eventually get invited to New York for more interviews and socializing. Those who succeed get an offer, and then get wined and dined in Manhattan at Arris' expense.

The task is no less demanding for current members at Arris, who must allocate time in their busy schedules to interview and socialize with candidates and convince them to join Arris rather than a competitor. The task is especially daunting with regard to hiring new associates (Reingold 1998). Within Arris, lists of candidates are maintained formally, along with the probability of their joining the firm. A team of people is assigned to monitor the candidate's inclinations toward Arris, answer any questions and concerns the candidate may have during the courting process, and generally to convince the candidate that Arris is a great choice.

For analysts, their experience at Arris is generally their first full-time, long-range (i.e., at least two years) job experience. Analysts are junior members of any client project; their work generally involves performing discrete modules or tasks under the supervision of a more experienced consultant or partner. "Analysts are professional members of [Arris] and take on a variety of responsibilities from carrying out data collection and essential research to conducting complex quantitative, financial and strategic analyses of businesses and corporations" (1998 Arris recruiting brochure). Analysts have several career paths, including staying with Arris through promotions, leaving Arris after two years to attend business school (and then returning, or not, to Arris as an associate), or leaving Arris to do something else. Analysts have varied educational backgrounds, in liberal arts, engineering, business, etc.

New associates are recent graduates of top M.B.A. programs and "in the early stages of their careers Associates are responsible for research and analysis as well as packaging findings" (1998 Arris recruiting brochure). Associates typically have prior work experience, usually in another field, but sometimes as former analysts at Arris who have accepted Arris' offer to pay for their graduate degrees and hire them upon graduation from the M.B.A. program.

New consultants are drawn to the work for several reasons. Many expressed the desire to be in an environment where they would learn continuously through their work, where they would be exposed to a variety of industries and client problems, where they would be surrounded by other bright people, and where they would learn valuable skills. They expect to be challenged to learn new industries and skills and are motivated by the prospect. Coming from top educational institutions, they expect to pick up many skills on their own, with minimal formal training. A couple of analysts expressed an alternative to this basic motivation. One indicated that the reason they were all there was for the salary that would allow them to pay off student loans and at the same time live a certain lifestyle in Manhattan. But many pointed out that they had chosen management consulting over investment banking (where salaries are generally higher) for a broader exposure to industries and skills, as well as to avoid the 100-hour work weeks (consultants only work 60, 70 or 80 hours!). Another indicated that while learning on the job was not an unusual demand, learning a new task and a new software tool late at night when you're exhausted, with a looming deadline, was perhaps expecting too much.

> I definitely learned, but it wasn't the best way to learn this. There was a lot of work, a lot of pressure, and I felt really poorly equipped to deal with it. Go run a query and make a pivot table is not a big deal. But when you're figuring out the finance, and trying how to figure it out in Excel, and then you're getting error messages because the data isn't flawless, it's a little bit much to handle, you know, at midnight, when you're getting really, really tired, and you're wondering why you took this job to begin with. [Interview Transcript]

7.2 Formal Training

All new analysts and associates (even those former analysts returning to the firm as associates) attend the same training sessions. New analysts and associates attend two mandatory training sessions. The first is a two-week session held in September when the majority of new hires begin their jobs. Each new consultant receives two thick binders covering the topics in training plus additional support materials. The September training covers "Orientation" training and "Tools of the Trade" training. Orientation introduces consultants to Arris and covers topics such as the Arris' growth and organization, employee benefits, travel and entertainment policies, staffing and mentoring, upward and downward employee review procedures, tips for traveling, and other administrative and practical advice. "Tools of the Trade" training covers functional areas, such as basic finance and accounting, and specific skills training, such as building spreadsheet models with Excel, conducting regression analysis, and when to use Access database software.

The second mandatory training session for new consultants, called "Project Simulation," occurs in January. This training occurs off-site, where new consultants from around the world meet at a five-star hotel to attend lectures, "solve" a simulated client project as part of a project team, and socialize together (see section 6.2).

7.3 Other Training, Other Mechanisms, Other People

In addition to the two formal training sessions provided for new consultants, other types of optional training are available. These include computer-based training (CBT), informal workshops offered periodically in the office, and informal workshop documents and tutorials available for down-loading from the server.

Beyond training and self-taught tutorials, consultants turn to others when they encounter something new or have a problem doing an analysis. While a few consultants are content to learn from books, manuals, or tutorials, the majority prefer to employ a strategy of asking others when the need-to-learn situation arises.

8. Discussion

8.1 Community of Practice at Arris

According to LPP, legitimacy occurs through productive work activities, and learning is dependent on newcomers' having legitimacy in the community and having access to members and resources of the community, rather than on formal training. At Arris, new consultants are engaged in productive work activities, i.e., assigned to a client team, very soon after joining the firm, with very little formal training. Further, they have wide access to other consultants and to resources such as information technology, and they are encouraged to seek out others and ask questions. Through engagement in productive work activities, consultants are expected to learn their work and how to use IT.

LPP suggests that newcomers will initially occupy less-engaged locations in the community. At Arris, this form of participation is manifest in new consultants' being assigned discrete tasks and being guided or directed in which IT to use to perform the task and how the IT may be used to perform the task. One experienced consultant explained a newcomer's role in these terms: "It's hard to strike a balance between telling someone exactly what you want and telling them basically where the case is going and allowing that person to think about it and come back with his first cut at it" (Interview Transcript).

8.2 The Role of Formal Training

The majority of new consultants interviewed stressed the importance of "learning-on-the-job" and learning from other people. While they recognized the importance of having a foundation, acquired either through their prior education or through training, they felt that the only way to really understand their jobs and tasks was by actually doing them. With respect to learning the software tools, one consultant explained,

> You just have to do it. You can read about it, but it's different. You have to just do it. It's helpful to read about it and have that as reference material, but you're the one who's going to have to do it at three in the morning. You should figure it out. [Interview Transcript]

Another recounted,

> I'm a firm believer that you can go through training, and you can sit at your computer and get familiar with how to use the package, but I think the learning is more experiential as opposed to something that can be taught. It's more learning-by-doing as opposed to memorizing or getting familiar with what you're supposed to be doing. [Interview transcript]

Most implementation initiatives of information systems include training users in how to use the system as an important element of the implementation. Training is often conceived of as occurring in a formal classroom setting, shortly before or after the system goes live. Some people have suggested continuous training for effective technology use (Strassmann 1990). Some recent research points to the ineffectiveness of traditional, formal training for knowledge workers and to the importance of peer networks for learning to use IT for effective work performance (George, Iacono and Kling 1995). Thus, although conventional wisdom advocates formal training programs for learning about performing jobs and how to use IT, this research demonstrates that formal training plays a limited, although important role, especially with respect to using IT for consulting work.

8.3 Using IT to Learn the Job and Using the Job to Learn IT

These knowledge workers recognize that they are continual learners. Many took the job for that reason. What they may not recognize is that at the same time they are learning how to use the information technology, the information technology is helping them to learn their jobs. There is a duality in learning. Most are confident that they will learn to use the technology when they need to. In this respect, they are using their jobs to learn about IT. While performing an economic analysis, an analyst will sharpen his skills in Excel, through practice, discovery, errors, frustration and asking others. At the same time, by using Excel, he may sharpen his problem-solving and problem-structuring skills, learning how to lay out the spreadsheet. One associate indicated how using the software helps to understand analytical tasks,

> I think that, or would hope that, the software program like Excel helps Analysts understand the analyses they are doing....For instance, you can plug numbers into a formula, and have it spit something out at you, but in order for that number to make sense, you have to think about how that formula is structured, and the inputs you put in. [Interview Transcript]

IT helps consultants learn their jobs in the conventional ways as well, through computer-based training, and by providing access to information.

This duality in learning could account for the tremendous difficulties experienced by some consultants who have limited backgrounds in using IT. They stumble over the IT in using it, do not understand why one program is more appropriate than another and cannot plan work, which will use IT and which is to be done by others. As a result they are less effective in performing their jobs. Happily, the majority of consultants come to the organization with a fair amount of experience using information technology. Some do not recognize this, but when pressed, they recalled programming courses they had taken in high school, software packages they had used in college and computer camp they had attended as children.

9. Conclusion

This paper presents early findings of an on-going interpretive case study of knowledge workers at a strategic management consulting firm. One of the benefits of ethnographic-style research is that it challenges the assumptions we hold (Myers 1999). The initial findings of the present research challenge several assumptions implicit in the IS literature, thereby suggesting avenues for future lines of inquiry. Theories of technology acceptance, such as TAM (Davis, Bagozzi and Warshaw 1989) and Task-Technology Fit (Goodhue 1995), and models of training (Bostrom, Olfman and Sein 1990) typically take an individualistic approach, giving little, if any, weight to group effects, yet the present research indicates that the community of practice has an important influence on these knowledge workers. These same theories typically consider a single information system or software program, yet the knowledge workers at Arris use a multitude of information systems and software programs. While the IT literature on training emphasizes formal, structured training, formal training is just one mechanism by which these knowledge workers learn their work and how to appropriate information technology. The knowledge workers of this study learn their jobs and how to appropriate IT from their peers, superiors and subordinates.

Future lines of inquiry could explore details of the community of practice. For example, what is the composition of the different communities of practice? Are there different communities of practice for learning different aspects of the job and information technology? Given that knowledge workers use a multitude of information technologies, one might ask, what are the implications for transfer of skills across information technologies? Future lines of inquiry could also address training issues, such as how can formal IT training be made effective? What structure should on-going training take?

The obvious limitation of this study is that it investigates a single firm and group of knowledge workers. Conclusions drawn here regarding the importance of informal, on-the-job learning may not generalize to other groups of knowledge workers at other management consulting firms or to different industries. For example, the quick minds of Arris consultants, their curiosity, and desire for challenge may account for their propensity to learn on-the-job rather than through formal training. Similarly, their highly outgoing personalities may explain their preference for learning from others around them.

In spite of the study's limitation, it offers valuable insight into the incorporation of information technology into the working lives of knowledge workers at one organiza-tion. Based on these early findings of this study, extant theories of technology use and acceptance could be modified to incorporate the complex dimensions of knowledge workers' jobs as discussed here, and as represented in Figure 1. For example, theories explaining individual user adoption of technology and user training may be broadened to account for the community of practice and the multitude of systems and functionality knowledge workers have at their disposal. Training research may be extended to include the effects of the community of practice and to consider the structure of an on-going professional development and information technology training program. Finally, in terms of reward systems, firms may need to modify their incentive policies to recognize and encourage the mentoring and information-sharing across the community of practice. Just as the adoption of groupware and knowledge management systems have incited firms to alter traditional reward structures to acknowledge individual knowledge-sharing (Davenport, Jarvenpaa and Beers 1996; Orlikowski 1996), so may firms need to encourage individual mentoring, sharing, and information-seeking behaviors in the work environment.

References

Agar, M. *The Professional Stranger.* San Diego, CA: Academic Press, Inc., 1996.

Barley, S. "The Alignment of Technology and Structure through Roles and Networks," *Administrative Science Quarterly* (35), 1990, pp. 61-103.

Bell, D. *The Coming of Post-Industrial Society.* New York: Basic Books, Inc., Publishers., 1973.

Bostrom, R. P.; Olfman, L.; and Sein, M. K. "The Importance of Learning Style in End-User Training," *MIS Quarterly,* March 1990, pp. 101-119.

Brown, J. S., and Duguid, P. "Organizational Learning and Communities-of-Practice: Toward a Unified View of Working, Learning, and Innovation," *Organization Science* (2:1), 1991, pp. 40-57.

Davenport, T. H.; Jarvenpaa, S. L.; and Beers, C. "Improving Knowledge Work Processes," *Sloan Management Review*, (Summer 1996, pp. 53-65.

Davis, F. D.; Bagozzi, R. P.; and Warshaw, P. R. "User Acceptance of Computer Technology: A Comparison of Two Theoretical Models," *Management Science*, (35:8), 1989, pp. 982-10003.

Davis, G. "Conceptual Model for Research on Knowledge Work," MISRC Working Paper, University of Minnesota., 1991.

DeLone, W. H., and McLean, E. R. "Information Systems Success: The Quest for the Dependent Variable," *Information Systems Research* (3:1), 1992, pp. 60-95.

DeLong, D. W. *"My Job is in the Box": A Field Study of Tasks, Roles, and the Structuring of Data Base-Centered Work*, Unpublished Dissertation, Boston University, Graduate School of Management, 1997.

Devin, P. "Porsche People and Ford Folks: Different Patterns of Using User-Modifiable Interfaces," Working Paper, Department of Information Systems. New York, New York University, 1994.

Drucker, P. F. *Post-Capitalist Society.* New York: Harper Collins Publishers, 1993.

Drucker, P. F. *Managing in a Time of Great Change.* New York: Truman Talley Books, 1995.

Fisher, K., and Fisher, M. D. *The Distributed Mind: Achieving High Performance through the Collective Intelligence of Knowledge Work Teams.* New York: Amacom, American Management Association, 1998.

Gallivan, M. J. "Value in Triangulation: An Analysis of Two Approaches for Combining Qualitative and Quantitative Methods," in *Qualitative Research in Information Systems,* A. S. Lee, J. Liebenau and J. DeGross (eds.). London: Chapman & Hall, 1997, pp. 417-444.

George, J. F.; Iacono, S.; and Kling, R. "Learning in Context: Extensively Computerized Work Groups as Communities-of-Practice," *Accounting, Management and Information Technology* (5:3/4), 1995, pp. 185-202.

Goodhue, D. "Understanding User Evaluations of Information Systems," *Management Science* (41:2), 1995, pp. 1827-1844.

Janesick, V. J. "The Dance of Qualitative Research Design: Metaphors, Methodolatry, and Meaning," *Strategies of Qualitative Inquiry,* N. K. Denzin and Y. S. Lincoln (eds.). Thousand Oaks, CA: Sage Publishing, 1998, pp. 35-55.

Klein, H. K., and Myers, M. D. "A Set of Principles for Conducting and Evaluating Interpretive Field Studies in Information Systems," *MIS Quarterly,* Special Issue on Intensive Research in Information Systems Using Qualitative, Interpretive, and Case Methods to Study Information Technology (23:1), 1999 forthcoming.

Kling, R.,and Scacchi, W. "The Web of Computing: Computer Technology as Social Organization," *Advances in Computers* (21:1), 1982, pp. 1-90.

Laudon, K. C., and Starbuck, W. H. "Knowledge and Information Work in Organizations," Working Paper, Stern School of Business, New York University, 1994.

Lave, J. *Cognition in Practice.* Cambridge, England: Cambridge University Press, 1988.

Lave, J., and Wenger, E. *Situated Learning: Legitimate Peripheral Participation.* Cambridge, England: Cambridge University Press, 1991.

Lucas, H. C., Jr. *The T-Form Organization.* San Francisco: Jossey-Bass Publishers, 1996.

Mankin, D.; Cohen, S. G.; and Bikson, T. K. *Teams and Technology: Fulfilling the Promise of the New Organization.* Boston: Harvard Business School Press, 1996.

Miles, M. B., and Huberman, A. M. *Qualitative Data Analysis.* Thousand Oaks, CA: Sage Publications, 1994.

Myers, M. "Ethnographic Research Methods in Information Systems," *IS World Net Virtual Meeting Center at Temple University,* March 8-11, 1999 (online). Available at http://interact.cis.temple.edu/~vmc(click on "guest").

Orlikowski, W. J. "The Duality of Technology: Rethinking the Concept of Technology in Organizations," *Organization Science* (3:3), 1992, pp. 398-427.

Orlikowski, W. J. "Improvising Organizational Transformation Over Time: A Situated Change Perspective," *Information Systems Research* (7:1), 1996, pp. 63-92.

Reingold, J. "And Now, Extreme Recruiting," *Business Week,* October 19, 1998, pp. 97-100.

Rogoff, B., and Lave, J. (eds.). *Everyday Cognition: Its Development and Social Context.* Cambridge, MA: Harvard University Press, 1984.

Rubin, H. J., and Rubin, I. S. *Qualitative Interviewing: The Art of Hearing Data.* Thousand Oaks, CA: Sage Publications, Inc., 1995.

Ruhleder, K.; Jordan, B.; and Elmes, M. B. "Wiring the 'New Organization': Integrating Collaborative Technologies and Team-Based Work," Annual Meeting of the Academy of Management, 1996.

Schultze, U. "A Confessional Account of an Ethnography About Knowledge Work," *MIS Quarterly*, Special Issue on Intensive Research in Information Systems Using Qualitative, Interpretive, and Case Methods to Study Information Technology, 1999.

Strassmann, P. A. *The Business Value of Computers*. New Canaan, CT: The Information Economics Press, 1990.

Strauss, A., and Corbin, J. *Basics of Qualitative Research*. Newbury Park, CA: Sage Publications, Inc., 1990.

Sulek, J., and Marucheck, A. "The Impact of Information Technology on Knowledge Workers—Deskilling or Intellectual Specialization?" *Work Study* (43:1), 1994, pp. 5-13.

Turner, J. A. "Computer Mediated Work: The Interplay Between Technology and Structured Jobs," *Communications of the ACM* (27:12), 1984, pp. 1210-1217.

Tyre, M. J., and Orlikowski, W. J. "Windows of Opportunity: Temporal Patterns of Technological Adaptation in Organizations," *Organization Science* (5:1), 1994, pp. 98-118.

Yin, R. K. *Case Study Research Design and Methods*. Thousand Oaks, CA: Sage Publications, 1988.

Zand, D. E. *The Leadership Triad*. New York: Oxford University Press, 1997.

About the Authors

Valerie Spitler is a candidate for the Ph.D. degree in the Information Systems Department at New York University's Stern School of Business. Valerie's research focuses on knowledge work and the role that information technology plays therein. After attending the 1997 IFIP WG8.2 conference on qualitative research in information systems, she decided to conduct qualitative research for her dissertation. She holds an M.B.A. from INSEAD (France) and a B.S. from the Wharton School.

Michael Gallivan is an Assistant Professor in the Computer Information Systems Department at the Robinson College of Business at Georgia State University in Atlanta. He holds a Ph.D. from the Sloan School of Management at MIT, and was an Visiting Assistant Professor at the Stern School of Business, New York University, prior to arriving at Georgia State University.. Mike's research focuses on the individual and organizational factors that influence implementation of technology innovations among both IT professionals as well as end users. In particular, he is interested in individual and group-level learning that occurs within organizations. Mike is currently investigating how organizations can develop sustainable competitive advantage through judicious use of outsourcing IT, and developing effective partnerships to manage such relationships. In addition to examining more IT outsourcing strategies, he is also examining interorganizational exchange relationships specifically analyzing the role of trust in virtual relationships across firm boundaries. He can be reached by e-mail at MGallivan@gsu.edu.

17 INCORPORATING SOCIAL TRANSFORMATION INTO THE INFORMATION SYSTEMS AND SOFTWARE DEVELOPMENT LIFECYCLES

Christopher J. Hemingway
Tom G. Gough
University of Leeds
United Kingdom

Abstract

All approaches to information systems and software development assume lifecycle models, which have a significant impact upon the perspective adopted for design and development. In recognition of the limitations of conventional software engineering lifecycles, the Information Systems (IS) community has focused upon gaining recognition of organizational and human issues in systems design. The Human-Computer Interaction (HCI) community has also tried to introduce user-centered practices into software engineering, but its success has been limited because it has augmented existing lifecycle models that are document or risk-driven. Yet, user-centered design can only be fully realized through the adoption of a user-centered model of the information systems lifecycle as the basis for developing ICT-based systems. This paper explains how several of the limitations of current IS and HCI theory and practice can be overcome by modeling development lifecycles in terms of social transformation. The paper then presents an information systems development lifecycle based upon social transformation and illustrates how a user-centered software development lifecycle can be integrated into the IS development process.

Keywords: Information systems development, systems lifecycle, software development, socio-cognitive theory of information systems, user-centered design.

1. Introduction

Owing to the overriding need to manage the complexity of software systems, conventional approaches to software engineering used technology-centered methods for design (see, for example, DeMarco 1979; Yourdon 1972). These approaches embody the "waterfall" model of the development lifecycle (Royce 1970), viewing analysis as the identification of the correct structure to encode the problem domain. Avison and Fitzgerald (1995) suggest that the software development community recognize the need for a more user-centered approach, although their development methods still leave design decisions under the control of technical experts. The empowerment of users during the development lifecycle has received attention by the Information Systems community for some years, recognizing the need for user participation in systems development and the importance of managing end-user computing. In light of progress in supporting user participation in major development efforts, Mumford (1983) proposed the following typology:

1. **Consultative participation**. Analysts discuss the system's requirements with users, but the technical experts perform design.
2. **Representative participation**. Representative users work on the design team and are, thereby, involved in decision making.
3. **Consensus participation**. Users drive the design process, making all key decisions.

An obvious trend in this list is the transfer of authority and responsibility for design decisions from technical experts to users. It has rarely been noted, however, that this transfer does not empower users unless they also have the *capacity* to act accordingly (see section 3 for a definition of this term). Furthermore, the education of users to improve their capacities for developing and using ICT-based information systems—necessary for the effective use of end-user computing in an organization—is poorly understood. Section 2 of this paper shows that, although existing approaches to systems development have considered authority, responsibility and capacity, their use of these concepts as the basis for guiding the development process has been limited. Furthermore, analyses of capacity are not acted upon during systems development to ensure that users have the information handling skills that the systems design implies. To address these weaknesses, this paper sets out a case for changes to information systems theory and practice through the adoption of user-centered information systems and software development lifecycles. The changes to IS theory, presented in section 3, are drawn from the socio-cognitive theory of information systems (Hemingway 1999; Hemingway and Gough 1998). Lifecycle models based upon this theory are developed in sections 4, 5 and 6 to demonstrate the effects of incorporating social transformation into information systems and software development. The implications for information systems and software engineering practice are then considered and conclusions drawn.

2. Current Perspectives on the Information Systems and Software Development Lifecycles

Software engineering differs from other engineering disciplines in two key respects:

- the physical environment is viewed in terms of peripheral constraints rather than as the basic medium for development; and
- engineers are primarily concerned with systems and pay comparatively little attention to the development of general systems components.

The consequence of these differences is that software requires a fundamentally different approach to engineering. Numerous methods have been proposed, but they all (with the exception of formal methods) typically result in software systems that have numerous design flaws. Quality management is being improved through the application of quality management techniques, such as the Capability Maturity Model (Ferguson and Sheard 1998; Herbsleb et al. 1997), and standards, such as ISO9001 and ISO15504. Nevertheless, the systemic approach to design and the absence of architectural principles for developing software systems limits the quality and reliability that can be achieved and contrasts sharply with, for example, electronic engineering, where applications are constructed from a very small number of precisely engineered modules (for example, standard amplifier circuits).

The most widely known model of the software development lifecycle is the conventional "waterfall" model (Royce 1970). Although this model is now regarded as grossly oversimplified, its influence on more recent lifecycle models is readily discerned. Two serious and well-documented deficiencies of such lifecycle models are considered here. First, the customer (who defines the system's requirements) and the users are excluded from much of the decision process. Consequently, many decisions are based upon the developer's interpretations of what the customer requires, supplemented by his or her understanding of what constitutes a user-friendly system. This limitation has been cited as a key reason for low levels of software acceptance (see, for example, Norman and Draper 1986). Second, despite the strong technical focus, the process fails to satisfy the technical criteria applied to other branches of engineering. Several alternatives to the waterfall model have been proposed as solutions to these and other development problems, the most widely used of which are incremental development, evolutionary development using prototypes, and rapid application development (RAD).

While incremental development represents a simple extension to the waterfall model, the use of prototyping in evolutionary development is a substantial change. A significant problem for many customers and users is that, although they can state their problem, they have insufficient experience of possible solutions to state their preferences between them and their contribution to design is, therefore, limited. Prototyping helps overcome this by providing users with experience of alternative software solutions and a mechanism for improving the dialogue between analyst, customer and user. Proto- typing must be carefully managed, however, to avoid slowing down development or unduly increasing costs and is only useful for studying some aspects of systems (Olle et al. 1988). A critical decision when using prototyping is whether to throw away the prototype or develop it into an operational system. Throwaway prototyping raises the prospect of long development times, as with the waterfall lifecycle, whereas developing the prototype risks poor software quality and increased maintenance costs.

Rapid application development extends prototyping tools to enable the production of workable applications for certain problem domains. RAD is particularly well-suited to transaction-based processes with well-defined inputs and relatively simple representations, such as tables, for output. From the customer's and users' perspectives, RAD has several potential drawbacks that must be carefully managed. Crucially, the focus on developing database systems that can readily be modified may encourage a short-term perspective and, thereby, shorten the time between revisions of the system. This has three potential consequences: (1) systems changes are made in response to user requests without full consideration of the strategic/organizational implications; (2) the coherence and integrity of the system may deteriorate more rapidly; and (3) savings at the initial development stage are offset by increased maintenance/redevelopment activity.

As illustrated above, software engineering has had limited success in improving the quantity and quality of customer and user participation. The HCI discipline, however, regards its primary goal as the integration of user-centered techniques into software engineering (Dix et al. 1993; Sutcliffe 1995). As illustrated by the example below, HCI methods tend to *augment* the software engineering process with user-centered tools and techniques, rather than *integrate* them at the methodological level. Consequently, development remains document or risk driven, albeit with an increased awareness of customer and user needs. The "psychological and organizational tools" developed by Clegg et al. (1996) are related to the waterfall model in Figure 1 to illustrate the limitations of augmenting software engineering approaches. With the waterfall model, the tools can provide only a list of tasks to be computerized and a method for filtering out unacceptable solutions *after* the development process. An evolutionary prototyping methodology would resolve some of these problems by allowing usability evaluation to feed back into the task allocation and job design processes, but this would require some revision of the tools and how they are collectively applied.

In comparison with software engineering and HCI, the information systems community has proposed more radical approaches to user involvement in systems development. Lyytinen (1987), for example, presents a taxonomy of information systems development methodologies based upon three contexts: technology, organization and language. The taxonomy is used to illustrate how the different premises of methodologies lead to different perceptions of the development process. A central point in Lyytinen's analysis is the demonstration that methodologies are partial in terms of their coverage of the three contexts. The socio-technical approaches, for example, emphasize the organization and technology contexts relative to the language context. In terms of developing a lifecycle model to underpin development methodologies, the three contexts are suitably accounted for by the notion of social transformation, as illustrated by the models presented in the following sections of this paper. The language context relates to the capacities of users to manipulate symbolic representations, which are the skills constitutive of information systems. The technology context refers to the competence for developing and manipulating technology artefacts in order to gain access to symbolic representations and to supplement language skills. The organizational context refers to the social relations that bear upon the access, control and change of information artefacts. A lifecycle model that uses social transformation as its central precept provides the potential for making methodologies contingent in their balancing of the three contexts proposed by Lyytinen. Given that systems development implies organizational change, a contingent approach is preferable to the use of a taxonomy to guide the selection of a methodology that is fixed in its treatment of social organization.

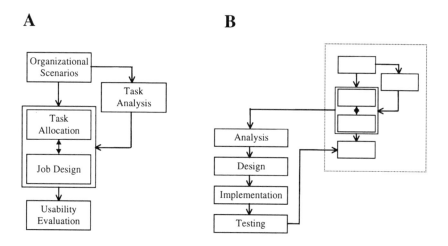

Figure 1. (A) Psychological and Organizational Tools (Reproduced from Clegg et al. [1996] courtesy of the Institute of Work Psychology, University of Sheffield, United Kingdom) and (B) Their Fit with the Waterfall Lifecycle

Iivari (1990a,1990b) develops a lifecycle model from a similar perspective to that adopted by Lyytinen. Extending Boehm's (1988) spiral model of the lifecycle and drawing upon the PIOCO development methodology, Iivari presents a three-phase evolutionary lifecycle. The three phases—organizational, technical and conceptual—are essentially synonmous with the three contexts proposed by Lyytinen. By incorporating these factors into a lifecycle model, Iivari addresses several of the problems identified by Lyytinen's taxonomy, resulting in a feasible, although somewhat complex, lifecycle model that is sensitive to social change. As illustrated above, Lyytinen's contexts and, hence, Iivari's three-phase approach focus on developing software rather than user skills. Thus, although Iivari recognizes that the evolutionary development involves a considerable amount of learning on the part of the developers, no explicit mechanism is proposed for ensuring that users participating in the development process have the requisite technical and conceptual skills. Furthermore, the lifecycle retains a focus on software and does not provide any specific process to support the development of users' information handling abilities.

Iivari, Hirschheim and Klein (1998) identify and compare five approaches to IS development that are significantly different than conventional software engineering. In a similar vein to Lyytinen's taxonomy, the analysis clarifies and contrasts the basic premises underlying the approaches. While this framework provides some useful insights, its analysis of ontological and epistemological positions has a number of weaknesses, the most significant being the failure to make a useful distinction between ontological and epistemological anti-realism (see Hacking 1983) and the narrow range of ontological commitments considered. The predominance of philosophical concerns in the analysis suggests that it is of academic, rather than practical, interest. Indeed,

Iivari, Hirschheim and Klein conclude with a number of comments about IS as an academic discipline. If extended to address the above weaknesses, however, the framework may prove a useful starting point for analyzing existing methodologies and considering how they might address issues of social transformation more thoroughly. Prerequisites for such methodological development are systems and software development lifecycles that regard social transformation as central to effective information systems development. The remainder of this paper describes such lifecycle models based upon the socio-cognitive theory of information systems.

3. Key Concepts from the Socio-cognitive Theory of Information Systems

The lifecycle models presented in the following sections draw upon the socio-cognitive theory of information systems (Hemingway 1999; Hemingway and Gough 1998). As the theory is only recently developed and not widely published, the concepts central to developing the models are outlined here. It is important to note that terms such as system and organization are used in the following sections according to their definitions in the socio-cognitive theory. The theory's definitions of agency, action, motivation and values are also used in developing the models, although their definitions are not provided in this paper.

3.1 Information Artefact

All communication takes place via some medium. Furthermore, some media can also store representations (i.e., arranged signs and symbols). Media capable of storage are classed as information artefacts and are characterized in terms of three modalities:
- **Modes of organization**: The use of, for example, spatio-temporal analogues or symbolic abstractions to encode a message into a medium.
- **Modes of selection**: The filtering of information prior to its representation.
- **Modes of navigation**: Mechanisms by which the user can interact with modes of organization and selection.

3.2 The Individual

When representing individual actors in an activity system, three factors are of the most interest: the individual's memories of their experiences as actors; the individual's interpretations of messages that he or she extracts from information artefacts; and the ability of the individual to act. The ability to act is dependent upon the physical relations between the individual's body and the environment and the individual's largely tacit knowledge of how to act in different situations. The individual's present knowledge of how to act defines his or her capacity for future action. The relationship between cognitive abilities, capacity and power are explained in detail in Hemingway (1999) and Hemingway and Gough (1999).

3.3 Social Organization

A distinguishing characteristic of the theory is its attempt to provide an integrated perspective on cognition, the body and communication. Thus, its analysis of social interactions begins with the consideration of physical co-presence, which is regarded as the basis for early forms of learning. Co-presence is important because it permits two or more individuals to refer to the same phenomena, even though they are all drawing upon their own experiences. Over time, the actions of co-present individuals can become regularized, thereby providing inter-subjective knowledge.

When considering social interaction within organizational forms, institutionalized practices and documented procedures and standards for action can be discerned (see, for example, Giddens 1984). It is important when intervening in social activity systems to have some understanding of the power relations that facilitate the maintenance of these regularized actions and standardized working practices. When considering information artefacts, the power issues can be considered in terms of (1) the actors' abilities to access information artefacts and interpret their message contents; (2) actors' abilities to control the message content of information artefacts; (3) authority over the access and control of information artefacts and their message contents; and (4) responsibility for the quality of information artefacts and their message contents.

4. Incorporating Social Transformation into the Information Systems Development Lifecycle

Social transformation is an intentional change to a social organization. In the context of information systems, such transformations must address six types of issues, which form the basis for any intentional information systems development:
1. the development of actors' capacities to represent, communicate and interpret messages and use the resulting knowledge to guide action;
2. the development of information artefacts to supplement the capacities identified in point 1;
3. the development of users' capacities to exploit ICT artefacts;
4. the provision of sufficient access to and control over information artefacts for the effective performance of tasks;
5. the alignment of power over with responsibility for information sources; and
6. the identification and resolution of deficiencies in the information available within the organizational form.

Information systems development begins when at least one stakeholder participating in an activity system perceives a deficiency in the system and subsequent analysis reveals that the deficiency can be resolved, at least in part, through changes to communication processes and information usage. Thus, as illustrated in Figure 2, the first two stages of the lifecycle are stakeholders' perceived dissatisfaction initiating the lifecycle, followed by the analysis of the activity system to determine whether the deficiency can suitably be addressed in information systems terms.

Where the contribution of IS developments is clearly identified, the next stage of analysis is to gain a detailed understanding of the stakeholders' perceived deficiencies and to classify these according to the six types listed above. The results of this analysis

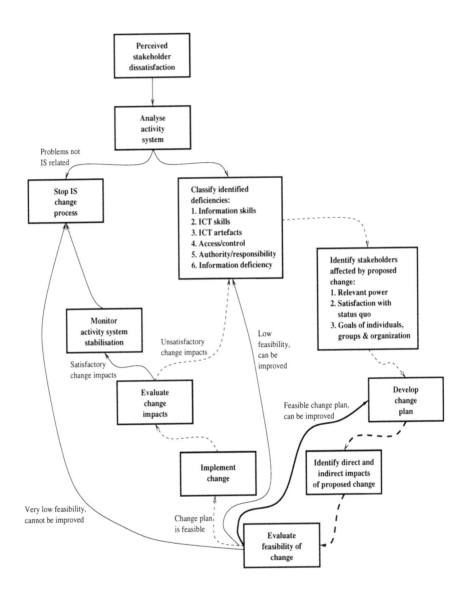

**Figure 2. An Information Systems Lifecycle Based Upon
the Socio-cognitive Theory of Information Systems**

of the activity system is the identification of activities and information artefacts that need
to be modified. Having identified ***what*** needs to be changed, the analysis goes on to
consider ***who*** will be affected by the changes. It should be noted at this stage that past
and present power relations and information needs will have led to the emergence of

stable information flows and standardized information artefacts within the activity system. Many such flows will be components of regularized working practices. Thus, unless widespread dissatisfaction is apparent, it can generally be assumed that the *status quo* serves the participants in the activity system reasonably well or, at least, satisfies those participants with most power. Consequently, a key issue to be addressed when identifying stakeholders affected by the proposed change is the assessment of stake-holders' power over the affected information flows and information artefacts and their satisfaction with the existing organization. The balance of satisfaction and power is a key factor determining the feasibility of any changes to the activity system. During this analysis, it may be useful to distinguish between the following stakeholder categories:

- **Current Information Users**: Those participants in the information system who are presently able to access or control an information artefact and/or its content.
- **Potential Information Users**: Those participants who have legitimate access to information artefacts, but presently do not have the capacity to manipulate the artefacts and/or interpret the messages that they contain.
- **Secondary Information Users**: Those participants in the activity system who have power to direct the activities of any current information users with respect to the use of the information flows and information artefacts in question.
- **Activity Relevant Stakeholders**: Those persons in the activity system who will be directly or indirectly affected by changes to the information flows and information artefacts being considered.

Given that stakeholders may themselves lack the capacities to make the requisite changes to the system, the following points should also be considered at this stage:

- Do stakeholders with the capacity and authority (i.e., legitimate power) to change information artefacts perceive a need for change?
- Do stakeholders with the authority to make the required changes perceive the need for changes but lack the capacity to make them?
- Do stakeholders with the capacity to make the changes lack the authority to enact them?

The analysis of stakeholders in the above terms will provide a good indication of which system changes are likely to be successfully initiated. A further consideration at this stage is the extent to which the changes affect the goals of stakeholders. The analysis of goals is one of the most difficult aspects of information systems analysis because the goals of individuals, groups and organizations must be considered, yet, as illustrated by the ontology of the socio-cognitive theory of information systems, all such goals ultimately reflect the motivations of, and interactions between, individual stakeholders.

Having determined what changes are to be made and who is likely to effect, affect and be affected by the changes, the next phase of development is to plan the change process. Owing to the diversity in goals and the complex network of power relations, the planning process takes the form of a negotiation cycle, illustrated by the bold lines in Figure 2. Accounting for the findings of the two analysis stages, an initial change plan is formulated. The direct and indirect impacts of the plan are then identified, perhaps including some consideration of the likely extent of "ripple effects" resulting from the changes made to the activity system. The anticipated impacts are then compared with the findings of the analysis to evaluate the feasibility of the planned changes. While an entire spectrum of findings may result from the evaluation, four broad types of outcome may result:

- the proposed changes are not implemented because the likelihood of success is low, the benefits are low and/or the associated risks are high;
- a second analysis phase is conducted because the proposed changes are found to be of low feasibility but are perceived to be feasible in the context of a change process that is essentially different in scope (i.e., it is realized at some stage that the perceived problem that triggered the lifecycle is not actually the central problem that needs to be addressed);
- the change plan is modified as a result of improvements identified during the evaluation of the initial plan; or
- the change plan is executed.

As illustrated by the negotiated implementation cycle (shown as a dashed line in Figure 2), it is not possible in practice to distinguish between the negotiation of the change plan and the implementation of change. The reason for this is that the precise nature of the change and its consequences cannot be completely predicted and, consequently, any actual change to an activity system must itself take place through a negotiation process. The negotiation of implementation is shown to include further analysis precisely because the impacts of the changes need to be understood and responded to if change is to be successful (analysis at this stage is not, of course, as detailed as during the initial stages of IS development). During the process of changing the activity system, new information artefacts may need to be introduced or existing artefacts redesigned. A guide for considering social transformation when developing information artefacts is presented in section 5. Where the information artefacts are ICT-based, a software development lifecycle, as described in section 6, will constitute part of the implementation cycle.

Following the negotiated changes to information artefacts and information usage, the activity system will eventually stabilize and negotiations will significantly reduce, or even cease. Provided that no other significant changes to the system occur, modified processes will eventually become regularized and accepted as social norms. It should be noted that the lifecycle described above has various iterative phases. Furthermore, numerous change processes can co-occur and interact, with the initial change process sometimes stimulating other changes. Such knock-on effects can even occur at the planning stage because the discussion of systems developments can affect the satisfaction of stakeholders with other aspects of the activity system. Although the change plan has effectively been completed at this stage, it is essential that monitoring of the changes takes place while the system settles into a new routine. If it becomes apparent that unexpected side-effects of the change are becoming problematic, a new information systems development lifecycle may be initiated.

5. Social Transformation and the Development of Information Artefacts

A key component of many information systems developments is the creation or modification of information artefacts. To provide a basis for the user-centered design of such artefacts, which includes software systems, Table 1 relates the key concepts of social transformation to three modalities that can be used to describe information artefacts.

Table 1. Socio-cognitive Concepts Used in the Design of Information Artefacts

Information Artefact	Individual	Social
Organization	Memory	Regularization and standardization
Selection	Interpretation	Communication
Navigation	Capacity	Power (Access, control, authority and responsibility)

Each row of Table 1 permits the characterization of information artefacts in their present and planned future states in terms of the social transformations they will support. Considering the first row, an information artefact can be regarded as located on a continuum. At one extreme, the artefacts are standardized to reflect well-established practices within the activity system (this is common in, for example, accounting systems). At the other extreme are flexible artefacts that can encode highly individualized messages. Typical examples include "notes" fields in structured databases and an employee's personal records about a client. Messages encoded in unstructured artefacts are subject to broader interpretation than standardized messages and are not always well-understood by individuals outside their immediate context of use.

Considering the second row of Table 1, information artefacts can be located on a continuum representing the encoding and extraction of messages. The possible means for encoding and extracting information are determined by the possible modes of organization. An artefact that encodes the details of numerous customers, for example, may be sorted in a particular order (or permit certain methods of sorting). The resulting organization affects the ways in which parts of the encoded message can be encoded and extracted, and the relative efficiency of encoding and extracting messages. Of most significance, however, is whether users interact *through* the information artefact or interact *with* it. The distinction between organization and selection can be subtle because, particularly with static artefacts, such as paper forms, the modes are closely coupled. Organization refers to what can be encoded by an information artefact and how. Selection refers to the means available to a user for manipulating a message during its interpretation.

The third row of Table 1 does not represent a continuum, but a number of factors that need to be considered to ensure the stability and effectiveness of the information system. Access to information artefacts relevant to a task is often critical to the effective performance of the task. Control refers to the ability of the user to modify the message content of an information artefact in order to communicate information to other users. Authority refers to the legitimate means by which a stakeholder may limit access to, and control over, information artefacts and their message contents. Responsibility for information artefacts and their message contents identifies the stakeholders who will be rewarded or sanctioned to reflect the quality of the information artefact and its message contents. Much of the analysis relevant to balancing these issues in a systems context are addressed by the IS development lifecycle. In terms of developing artefacts, the main concern is providing modes of navigation by which information access and control can be managed and, particularly in terms of software systems, authority and responsibility implemented.

6. Incorporating Social Transformation into the Software Development Lifecycle

As illustrated in section 4, information systems development addresses six interrelated concerns. The information systems community has typically focused on issues three to six, with comparatively little attention paid to skills issues during the development process. Although the authors stress that most benefits will arise from IS developments that address all six issues and their interrelationships, this section extends the IS development lifecycle to propose an alternative to the software development lifecycles considered in section 2.

The software development lifecycle is an abstraction of specific software concerns from the more general IS development lifecycle. This abstraction can be achieved with the above lifecycle by identifying the ICT artefact requirements that can be effectively supported by a software solution. The decision to develop computer-based information artefacts must take into account the implications particularly in terms of required ICT skills, but also in terms of "information ownership" and other power-related issues that are affected by the properties of information artefacts. The analysis presented in section 5 provides a framework for appraising the likely impacts of information artefacts on power relations in an activity system, which can be used as a guide for software design. The use of Table 1 as a framework is quite readily achieved because the description of information artefacts in terms of organization, selection and navigation relates quite directly to concepts used in conventional data modeling and software design. Modes of organization, for example, correspond with the logical data model in terms of data storage and human-computer interface design. Modes of selection correspond with both user and system operations for encoding, sorting and filtering. Modes of navigation refer to the dialogue between human and computer and, at the systems level, to the logical arrangement of data files and the interconnections between them that are achievable with the operating system and applications available. Further research needs to be conducted before a detailed software lifecycle and development methodology can be produced. An indication of future work by the authors is presented in section 8.

7. Conclusions

Only limited success has followed the attempts by both the software engineering, IS and HCI communities to change information systems and software development practices to make them more user-centered and more sensitive to the organizational and human context in which development takes place. The reason for such limited success is that user-centered design can only be fully realized through the adoption of user-centered lifecycle models on which to base development processes.

User-centered systems and software development need to be integral to information systems theory and practice. This end can be achieved by incorporating social transformation into these lifecycles, which is made possible by the socio-cognitive theory of information systems.

8. Future Work

Future research issues relevant to the further development of the socio-cognitive theory of information systems and the lifecycles presented in this paper, which the authors aim to address, are:

- the study of information skills and how they can be developed in users;
- the relationships between users' information skills, task characteristics, the properties of information artefacts and task performance;
- ways of improving user capacities for developing information artefacts and information systems to support their tasks;
- the negotiation of development capacities between IS and software engineering professionals and users;
- the development of detailed IS and software development methodologies based upon the lifecycle models presented in this paper; and
- the evaluation of alternative software solutions in terms of their information and IS development skills requirements.

References

Avison, D. E., and Fitzgerald, G. *Information Systems Development: Methodologies, Techniques and Tools*, 2nd ed. Maidenhead, England: McGraw-Hill Publishing Company, 1995.

Boehm, B. W. "A Spiral Model of Software Development and Enhancement," *Computer*, 1988, pp. 61-72.

Clegg, C.; Coleman, P.; Hornby, P.; Maclaren, R.; Robson, J.; Carey, N.; and Symon, G. "Tools to Incorporate Some Psychological and Organizational Issues During the Development of Computer-based Systems," *Ergonomics* (39:3), 1996, pp. 482-511.

DeMarco, T. *Structured Analysis and Systems Specification.* London: Prentice-Hall, 1979.

Dix, A.; Findlay, J.; Abowd, G.; and Beale, R. *Human-computer Interaction.* Englewood Cliffs, NJ: Prentice-Hall, 1993.

Ferguson, J., and Sheard, S. "Leveraging Your CMM Efforts for IEEE/EIA 12207," *IEEE Software*, September/October 1998, pp. 23-28.

Giddens, A. *The Constitution of Society.* Cambridge, MA: Polity Press, 1984.

Hacking, I. *Representing and Intervening: Introductory Topics in the Philosophy of Natural Science.* Cambridge, England: Cambridge University Press, 1983.

Hemingway, C. J. "Toward a Socio-cognitive Theory of Information Systems: An Analysis of Key Philosophical and Conceptual Issues," in *Information Systems: Current Issues and Future Changes*, T. J. Larsen, L. Levine, and J. I. DeGross (eds.). Laxenburg, Austria: International Federation for Information Processing, 1999, pp. 275-286.

Hemingway, C. J., and Gough, T. G. "A Socio-cognitive Theory of Information Systems," Technical Report 98.25, School of Computer Studies, University of Leeds, United Kingdom, December 1998.

Herbsleb, J.; Zubrow, D.; Goldenson, D.; Hayes, W.; and Paulk, M. "Software Quality and the Capability Maturity Model," *Communications of the ACM* (40:6), 1997, pp. 30-40.

Iivari, J. "Hierarchical Spiral Model for Information System and Software Development Part 1: Theoretical Background," *Information and Software Technology* (32:6), 1990a, pp. 386-399.

Iivari, J. "Hierarchical Spiral Model for Information System and Software Development Part 2: Design Process," *Information and Software Technology* (32:7), 1990b, pp. 450-458.

Iivari, J.; Hirschheim, R.; and Klein, H. K. "A Paradigmatic Analysis Contrasting Information Systems Development Approaches and Methodologies," *Information Systems Research* (9:2), 1998, pp. 164-193.

Lyytinen, K. "A Taxonomic Perspective of Information Systems Development: Theoretical Constructs and Recommendations," Chapter 1 in *Critical Issues in Information Systems Research*, R. J. Boland and R. A. Hirschheim (eds.). Chichester, England: John Wiley and Sons, 1987.

Mumford, E. *Designing Participatively: Participative Approach to Computer System Design*. Manchester, England: Manchester Business School, 1983.

Norman, D. A., and Draper, S. W. *User Centered Systems Design: New Perspectives on Human-computer Interaction*. Hillsdale, NJ: Lawrence Erlbaum Associates, 1986.

Olle, T. W.; Hagelstein, J.; Macdonald, I. G.; Rolland, C.; Sol, H. G.; van Assche, F. J. M.; and Verrijn-Stuart, A. A. *Information Systems Methodologies: A Framework for Understanding*. Wokingham, England: Addison-Wesley Publishing Company, 1988.

Royce, W. W. "Managing the Development of Large Software Systems," in *Proceedings of WESTCON*, San Francisco, CA, 1970.

Sutcliffe, A. *Human-computer Interface Design* 2nd ed. Basingstoke, England: MacMillan, 1995.

Yourdon, E. *Design of On-line Computer Systems*. Englewood Cliffs, NJ: Prentice-Hall, 1972.

About the Authors

Christopher Hemingway is with the Information Systems Research Group, School of Computer Studies, at the University of Leeds, United Kingdom. His main research interests are user-centered systems and software development; systems and software evaluation; and computer support for the modeling of information by experts and the communication of information between experts and non-experts. His research provides an integrated treatment of these issues through the development of the socio-cognitive theory of information systems. He can be reached by e-mail at cjh@ieee.org.

Tom Gough is also with the Information Systems Research Group, School of Computer Studies, at the University of Leeds, United Kingdom. His research interests relate to the "why" and "how" of information systems and include user-centered development and co-developing the socio-cognitive theory. Tom also has a general interest in medical informatics, particularly addressing health and safety issues during in IS development. He can be reached by e-mail at tgg@scs.leeds.ac.uk.

Part 3:

Panels

18 THE USES AND ABUSES OF EVALUATIVE CRITERIA FOR QUALITATIVE RESEARCH METHODS

Richard Baskerville
Georgia State University
U.S.A.

Steve Sawyer
Syracuse University
U.S.A.

Eileen Trauth
Northeastern University
U.S.A.

Duane Truex
Georgia State University
U.S.A.

Cathy Urquhart
University of the Sunshine Coast
Australia

The focus of this panel is on the impact of recent trends toward the establishment of quality criteria for evaluating qualitative information systems research. Advances in the use of qualitative and intensive research methods in information systems have raised debates about the methods for the evaluation of this form of research. Such criteria are not new in the general social science literature (e.g., Gummesson 1988; Kirk and Miller 1986; Yin 1989). However, these criteria have been more generally directed toward objective forms of qualitative research than toward interpretive forms of qualitative research. More recent work within the organizational (Golden-Biddle and Locke 1993) and information systems fields (Klein and Myers 1999) may provide general criteria for evaluating interpretive research. In addressing the issues that arise from the availability of these criteria, the panel will be structured with approximately half the time period dedicated to an opening discussion by the panelists, followed by an open forum debate engaging the audience to occupy the remaining half of the time period.

Richard Baskerville will open the panel with a brief overview of the current state of criteria used for evaluating qualitative research in information systems. Following this overview, Baskerville will summarize the advantages and disadvantages of criteria with regard to the quality goals of journals and conference venues.

Steve Sawyer will discuss how quality criteria affect the use of multi-method approaches to IS/IT research. Multi-method approaches often mean combining data collection and analysis techniques which have differing philosophical and analytic traditions (such as a combination of surveys and fieldwork). Further, while there is continual encouragement to employ multiple methods, there are relatively few studies of this nature in the IS/IT literature. Drawing on an ongoing multi-method study as an example (and not as an exemplar!), he will outline how general guidelines for conducting multi-method studies are enacted. He will also discuss what sorts of issues and implications arise from design decisions driven, in part, by the recognition of guidelines.

Eileen Trauth will consider the debate from the perspective of a researcher. She will discuss the practical consequences of having an established set of quality criteria on the conduct of qualitative (and in particular, interpretive) research. She will present an argument in favor of having such criteria. Her arguments will focus on the politics of publishing (both journal articles and books), faculty reward systems, and the allocation of energy and effort (on the research itself or on justifying it). She will express the viewpoint that establishing criteria need not be a constraining act but, rather, the beginning of a dialectical process which will constantly change (and hopefully improve) the conduct of qualitative research.

Duane Truex will review the impact of qualitative criteria on the emergence of the IS field. In so doing, he will be taking the position of the loyal opposition and will offer a provocative and cautionary deconstruction of a standard and thus challenge the position that a semblance of fixed standards are required at all. He will begin with a brief genealogy of the discourse in which qualitative research was introduced and finally gained acceptance in our research community. He will then offer a suggestion as to how the field might concern itself more with the social and continuous process of standardizing rather than adjudicating strictures. That is, the question of standards might be better viewed as a continuous and emergent language game.

Cathy Urquhart will discuss the principles from the perspective of having recently applied them to evaluate a recent project that encompassed hermeneutic study of texts and also used grounded theory. In particular, she will discuss the utility of such principles, how they represent an important debating point, and whether indeed such principles can be applied without becoming an orthodoxy that goes against the pluralistic nature of interpretive research.

References

Golden-Biddle, K., and Locke, K. "Appealing Work: An Investigation of How Ethnographic Texts Convince," *Organization Science* (4:4), 1993, pp. 595-616.

Gummesson, E. *Qualitative Methods in Management Research.* Lund, Sweden: Studentlitteratur Chartwell-Bratt, 1988.

Kirk, J., and Miller, M. L. *Reliability and Validity in Qualitative Research.* Newbury Park, CA: Sage, 1986.

Klein, H. K., and Myers, M. D. "A Set of Principles for Conducting and Evaluating Interpretive Field Studies in Information Systems," *MIS Quarterly* (23:1), March 1999, pp. 67-93.

Yin, R. *Case Study Research: Design and Methods.* Newbury Park, CA: Sage, 1989.

19 RE-EVALUATING POWER IN INFORMATION RICH ORGANIZATIONS: NEW THEORIES AND APPROACHES

Ole Hanseth
University of Oslo
Norway

Susan V. Scott
London School of Economics
and Political Science
United Kingdom

Leiser Silva
University of Alberta
Canada

Edgar A. Whitley
London School of Economics
and Political Science
United Kingdom

1. Introduction

The widespread use of innovative approaches to studying information systems in their organizational contexts is increasing our knowledge of the diverse, complex world-building activities of the various actors and stakeholders associated with the system. Unfortunately, in so doing, it is also demonstrating the inadequate nature of many of our theoretical constructs for understanding and explaining these activities. We have a richer picture of what is going on but are unable to match it with a more sophisticated mechanism for explaining it and so can only draw limited implications for theory and practice.

The aim of this panel is to introduce a range of new perspectives that can help information systems researchers in their tasks, focusing in particular on questions of power. Although the range of perspectives is broad and the individual positions may not be consistent with one another, they share a common theme in that they all draw on recent developments from philosophy and social theory, which the presenters feel provide the richest resources for understanding the role and impact of information and communication technologies in information rich organizations.

2. Format

Each of the presenters will briefly introduce a particular theoretical approach explicitly addressing how their chosen approach addresses questions of understanding power relations and explaining the complex phenomena that arise from the introduction and use of new technologies. They will then discuss the implications for empirical work that arise from the approach, drawing on recent studies that have been undertaken.

It is intended that plenty of time will be set aside for questions and discussion. To support this aim a document containing key readings on the theories presented and publications demonstrating the application of the theories will be made available to participants. A copy of the document will be available online at: http://is.lse.ac.uk/whitley/publications/pdf/powerpanel.pdf.

3. Contributions

Ole Hanseth will talk about Heidegger's notion of *gestell* and will demonstrate how it helps explain the new management structures and shifting power relations that underlie large infrastructure projects.

Leiser Silva will introduce Clegg's Circuits of Power model. He will then show how it can be used to integrate different views of power in the context of a particular information systems case study.

Edgar Whitley will then focus on one of these Circuits and introduce Bruno Latour's ideas about the power of human and non–human stakeholders in the context of socio-technical collectives. He will propose a new mechanism for considering all these stakeholders in the process of systems development.

Finally, Susan Scott will illustrate her study of the changing role of middle managers by drawing on Ulrich Beck's notion of the risk society, which provides a complementary perspective to more traditional notions of power.

Index of Contributors